Activities Manual for
Communication Electronics

Louis E. Frenzel, Jr.

GLENCOE
Macmillan/McGraw-Hill

New York, New York
Columbus, Ohio
Mission Hills, California
Peoria, Illinois

ACKNOWLEDGMENTS

The *Basic Skills in Electricity and Electronics* series was conceived and developed through the talents and energies of many individuals and organizations.

The original, on-site classroom testing of the texts and manuals in this series was conducted at the Burr D. Coe Vocational Technical High School, East Brunswick, New Jersey; Chantilly Secondary School, Chantilly, Virginia; Nashoba Valley Technical High School, Westford, Massachusetts; Platt Regional Vocational Technical High School, Milford, Connecticut; and the Edgar Thomson, Irvin Works of the United States Steel Corporation, Dravosburg, Pennsylvania. Postpublication testing took place at the Alhambra High School, Phoenix, Arizona; St. Helena High School, St. Helena, California; and Addison Trail High School, Addison, Illinois.

Early in the publication life of this series, the appellation "Rainbow Books" was used. The name stuck and has become a point of identification ever since.

In the years since the publication of this series, extensive follow-up studies and research have been conducted. Thousands of instructors, students, school administrators, and industrial trainers have shared their experiences and suggestions with the authors and publishers. To each of these people we extend our thanks and appreciation.

Activities Manual for Communications Electronics

Imprint 1992
Copyright © 1990 by the Glencoe Division of Macmillan/McGraw-Hill School Publishing Company. All rights reserved. Printed in the United States of America. Except as permitted under the United States Copyright Act of 1976, no part of this publication may be reproduced or distributed in any form or by any means, or stored in a database or retrieval system, without the prior written permission of the publisher.

Send all inquiries to:
GLENCOE DIVISION
Macmillan/McGraw-Hill
936 Eastwind Drive
Westerville, Ohio 43081

ISBN 0-07-058231-9

2 3 4 5 6 7 8 9 10 11 12 13 14 15 MAL 00 99 98 97 96 95 94 93 92

Contents

Editor's Foreword

The McGraw-Hill *Basic Skills in Electricity and Electronics* series has been designed to provide entry-level competencies in a wide range of occupations in the electric and electronic fields. The series consists of coordinated instructional materials designed especially for the career-oriented student. Each major subject area covered in the series is supported by a textbook, an activities manual, and a teacher's manual. All the materials focus on theory, practices, applications, and experiences necessary for those preparing to enter technical careers.

There are two fundamental considerations in the preparation of materials for such a series: the need of the learner and the needs of the employer. The materials in this series meet these needs in an expert fashion. The authors and editors have drawn upon their broad teaching and technical experiences to accurately interpret and meet the needs of the student. The needs of business and industry have been identified through questionnaires, surveys, personal interviews, industry publications, government occupational trend reports, and field studies.

The processes used to produce and refine the series have been ongoing. Technological change is rapid and the content has been revised to focus on current trends. Refinements in pedagogy have been defined and implemented based on classroom testing and feedback from students and teachers using the series. Every effort has been made to offer the best possible learning materials.

The widespread acceptance of the *Basic Skills in Electricity and Electronics* series and the positive responses from users confirm the basic soundness in the content and design of these materials as well as their effectiveness as learning tools. Teachers will find the texts and manuals in each of the subject areas logically structured, well-paced, and developed around a framework of modern objectives. Students will find the materials readable, lucidly illustrated, and interesting. They will also find a generous amount of self-study and review materials to help them determine their own progress.

The publisher and editor welcome comments and suggestions from teachers and students using the materials in this series.

Charles A. Schuler
Project Editor

BASIC SKILLS IN ELECTRICITY AND ELECTRONICS
Charles A. Schuler, Project Editor

Books in this series:

Introduction to Television Servicing by Wayne C. Brandenburg
Electricity: Principles and Applications by Richard J. Fowler
Communication Electronics by Louis E. Frenzel, Jr.
Instruments and Measurements by Charles M. Gilmore
Microprocessors: Principles and Applications by Charles M. Gilmore
Small Appliance Repair by Phyllis Palmore and Nevin E. André
Electronics: Principles and Applications by Charles A. Schuler
Digital Electronics by Roger L. Tokheim

Preface

This activities manual is designed to accompany the text *Communication Electronics*. It provides a variety of projects and activities to supplement and reinforce the contents of the text. It helps the student to review the key points discussed and, where possible, to verify the theories presented.

For each chapter in the text, there is a test in this manual covering the major topics. The test may serve as the primary chapter exam or as a review for another exam. Such exams not only provide a way to measure student comprehension, they also serve to verify that the learning objectives for the chapter have been met. Further, exams are an excellent learning activity that reinforces the important concepts covered in the text.

Most of the activities are lab experiments designed to put theory into practice. Students will build real communications circuits and will test them and use them. Such "hands-on" experiments provide real-world experience in the use of components, circuits, and test equipment.

Each experiment requires that a circuit be built. The circuit must then be tested for proper operation. Next, the experiment is run, following step-by-step procedures. This helps the student collect relevant data, draw conclu-

sions, and observe actual circuit operation and characteristics. Along the way, the student learns the operation of test instruments, special measurement techniques, and in many cases, troubleshooting methods to make the circuit operational.

Most of the experiments can be built with standard, readily available components on existing breadboards or trainers. No special components are needed. Existing lab power supplies, function generators, oscilloscopes, and multimeters may also be used. The only special instrument needed is a spectrum analyzer, and that can be rented when needed.

Although most communications circuits operate at radio frequency and microwave frequency, those used in these lab experiments are designed to operate at audio and subradio frequencies. Radio frequency signals and circuits are difficult to work with. Circuit layouts are critical, and component leads must be kept short in practice to prevent circuit oscillation and other problems. This is not practical in a lab situation. The low-frequency circuits are less critical and more forgiving in a lab setting. The concepts are more easily demonstrated at the low frequencies than at the high frequencies. Radiation problems and interference from multiple circuits in the lab are also eliminated by using lower frequencies.

Each experiment in this manual has been performed in a classroom environment, and the results verified. The author welcomes feedback from users on their experiences with this manual.

Louis E. Frenzel

Safety

Electric and electronic circuits can be dangerous. Safe practices are necessary to prevent electrical shock, fires, explosions, mechanical damage, and injuries resulting from the improper use of tools.

Perhaps the greatest hazard is electrical shock. A current through the human body in excess of 10 milliamperes can paralyze the victim and make it impossible to let go of a "live" conductor or component. Ten milliamperes is a rather small amount of electric flow: it is only *ten one-thousandths* of an ampere. An ordinary flashlight uses more than 100 times that amount of current!

Flashlight cells and batteries are safe to handle because the resistance of human skin is normally high enough to keep the current flow very small. For example, touching an ordinary 1.5-V cell produces a current flow in the microampere range (a microampere is one-millionth of an ampere). This much current is too small to be noticed.

High voltage, on the other hand, can force enough current through the skin to produce a shock. If the current approaches 100 milliamperes or more, the shock can be fatal. Thus, the danger of shock increases with voltage. Those who work with high voltage must be properly trained and equipped.

When human skin is moist or cut, its resistance to the flow of electricity can drop drastically. When this happens, even moderate voltages may cause a serious shock. Experienced technicians know this and they also know that so-called low-voltage equipment may have a high-voltage section or two. In other words, they do not practice two methods of working with circuits: one for high voltage and one for low voltage. They follow safe procedures at all times. They do not assume protective devices are working. They do not assume a circuit is off even though the switch is in the OFF position. They know the switch could be defective.

As your knowledge and experience grows, you will learn many specific safe procedures for dealing with electricity. In the meantime:

1. Always follow procedures.
2. Use service manuals as often as possible. They often contain specific safety information.
3. Investigate before you act.
4. When in doubt, *do not act*. Ask your instructor or supervisor.

GENERAL SAFETY RULES FOR ELECTRICITY AND ELECTRONICS

Safe practices will protect you and your fellow workers. Study the following rules. Discuss them with others, and ask your instructor about any that you do not understand.

1. Do not work when you are tired or taking medicine that makes you drowsy.
2. Do not work in poor light.
3. Do not work in damp areas or with wet shoes or clothing.
4. Use approved tools, equipment, and protective devices.
5. Avoid wearing rings, bracelets, and similar metal items when working around exposed electric circuits.
6. Never assume that a circuit is off. Double check it with an instrument that you are sure is operational.
7. Some situations require a "buddy system" to guarantee that power will not be turned on while a technician is still working on a circuit.
8. Never tamper with or try to override safety devices such as an interlock (a type of switch that automatically removes power when a door is opened or a panel removed).
9. Keep tools and test equipment clean and in good working condition. Replace insulated probes and leads at the first sign of deterioration.
10. Some devices, such as capacitors, can store a *lethal* charge. They may store this charge for long periods of time. You must be certain these devices are discharged before working around them.
11. Do not remove grounds, and do not use adaptors that defeat the equipment ground.
12. Use only an approved fire extinguisher for electric and electronic equipment. Water can conduct electricity and may severely damage equipment. Carbon dioxide (CO_2) or halogenated-type extinguishers are usually preferred. Foam-type extinguishers may also be desired in some cases. Commercial fire extinguishers are rated for the type of fires for which they are effective. Use only those rated for the proper working conditions.
13. Follow directions when using solvents and other chemicals. They may be toxic, flammable, or may damage certain materials such as plastics.
14. A few materials used in electronic equipment are toxic. Examples include tantalum capacitors and beryllium oxide transistor cases. These devices should not be crushed or abraded, and you should wash your hands thoroughly after handling them. Other materials (such as

heat shrink tubing) may produced irritating fumes if overheated.

15. Certain circuit components affect the safe performance of equipment and systems. Use only exact or approved replacement parts.

16. Use protective clothing and safety glasses when handling high-vacuum devices such as picture tubes and cathode ray tubes.

17. Don't work on equipment before you know proper procedures and are aware of any potential safety hazards.

18. Many accidents have been caused by people rushing and cutting corners. Take the time required to protect yourself and others. Running, horseplay, and practical jokes are strictly forbidden in shops and laboratories. Circuits and equipment must be treated with respect. Learn how they work and the proper way of working on them. Always practice safety; your health and life depend on it.

CHAPTER | 1

Introduction to Electronic Communications

ACTIVITY 1-1
PROJECT: COMMUNICATIONS METHODS

Electronic communications is so common today that we all take it for granted. Almost daily, we all use multiple electronic communications services.

To gain a better appreciation for the convenience and benefits that electronic communications bring us, consider how humans communicated over short and long distance before the telegraph, telephone, and radio were invented. Then do the following:

1. List all the ways you can think of that humans communicated at a distance without electrical or electronic apparatus. Include those given in the text as well as any of your own. Be specific.
2. List all the electronic communications methods you yourself use on a daily, weekly, monthly, and annual basis.

ACTIVITY 1-2
PROJECT: COMMUNICATIONS APPLICATIONS

To increase your knowledge of the scope and impact of electronic communications, list 10 applications not specifically mentioned in the text. They can be variations or subcategories of those in the text or entirely different ones.

ACTIVITY 1-3
PROJECT: LISTENING TO BC, TV, SW, AND VHF/UHF

Monitor the electromagnetic frequency spectrum to get a feel for the volume and variety of communications taking place.

1. At night (only), slowly tune the entire AM radio broadcast band from 550 to 1650 kHz. A car radio with antenna fully extended is good for

this purpose. Count the total number of stations you hear. Listen specifically to the very weak signals, and listen long enough to get a call sign or location. Record the cities or stations heard farthest from you. Repeat during daylight hours.

2. Switch your TV to all available channels from 2 to 83. List the local channels you normally watch, but search for weaker farther-away stations. Do this several times over a period of weeks at different times of the day. Record all channels and stations received, including the weak signals. List the most remote station.

3. Borrow or purchase a shortwave radio. Tune the frequencies from 2 to 30 MHz and look for the following:
    a. Marine communications on 2182 kHz
    b. Morse code (CW) signals*
    c. Time signals from station WWV on 5, 10, 15, or 20 MHz
    d. Ham radio signals on 3.5–4, 7–7.3, 14–14.3, 21–21.5, and 28–30 MHz*
    e. Foreign broadcasts

4. Borrow or purchase a scanner radio for listening to FM two-way communications channels in the VHF/UHF range. Listen to and try to identify the following:
    a. Police, highway patrol, and sheriff
    b. Dispatchers (taxi, truck, etc.)
    c. National Oceangraphic and Atmospheric Administration (NOAA) weather radio channels on 162 MHz

ACTIVITY 1-4
PROJECT: EXPERIENCING
RADIO COMMUNICATIONS

Experience different forms of electronic communications yourself.

1. Borrow or purchase a cordless telephone. Use it to make a call.
2. Borrow or purchase a CB radio. Listen to all 40 channels, especially the emergency channel, 9. Participate in CB by making contact with at least two stations. (No license is required.)
3. Ask around and identify a local ham radio operator in your school or neighborhood. Ask him or her to give you a demonstration of ham radio communications.
4. Locate an individual who uses a personal computer for communicating with a remote computer by using a modem. Ask for a demonstration.

*Use the beat frequency oscillator (BFO) on the receiver to receive code (CW) and SSB voice signals.

CHAPTER | 2

Amplitude and Single-Sideband Modulation

ACTIVITY 2-1
TEST: AMPLITUDE AND SINGLE-SIDEBAND MODULATION

Read the question and write the answer on a separate sheet of paper.

1. (True or False) In AM, the frequency of the carrier varies with the modulating signal. _____

2. (True or False) Only sine wave intelligence signals can be used in AM. _____

3. An amplitude modulator must perform the mathematical operation of _____.

4. (True or False) SSB is a type of AM. _____

5. An AM system has a 1.5-V carrier and a 0.8-V intelligence signal. The modulation index $m =$ _____.

6. An AM signal displayed on an oscilloscope has a peak-to-peak minimum trough voltage of 6 mV and a peak-to-peak maximum voltage of 24 mV. The percent modulation is _____.

7. Overmodulation occurs when: _____
   a. $V_m = V_c$
   b. $V_m < V_c$
   c. $V_m > V_c$

8. A 2.182-MHz carrier has two modulating tones of 800 Hz and 1.2 kHz. The sidebands are _____.

9. What is the total bandwidth of the AM signal described in question 8? _____

10. In a 100 percent modulated AM signal, _____ percent of the total power is in the sidebands.

11. A carrier of 50 W is 80 percent modulated. The power in one sideband is _____ watts.

12. It takes _____ W of audio power to 100 percent modulate a 2.5-kW carrier.

13. Overmodulation causes the bandwidth of an AM signal to
_____.
    a. Increase   c. Remain the same
    b. Decrease

14. AM without the carrier is called _____.

15. The primary advantage of SSB is _____.

16. An SSB transmitter produces 265-V peak-to-peak voltage across a
50-Ω antenna on voice peaks. The PEP is _____ W. The average
power output is _____ W.

17. (True or False) Mixing is the same as AM. _____

18. Another name for mixing is _____.

19. A mixer has 1.1-GHz and 750-MHz inputs. The outputs are
_____.

20. (True of False) Mixer inputs must not be modulated. _____

ACTIVITY 2-2
LAB EXPERIMENT:
MEASURING THE PERCENT
OF MODULATION

PURPOSE

To learn two methods of expressing and measuring the percent of modulation of an AM signal and to observe the condition of overmodulation.

MATERIALS

Qty.
1 Oscilloscope with provision for external horizontal input
1 Frequency counter
1 12-V dc power supply
1 Function generator with sine output
1 XR-2206 Exar function generator IC
1 0.002-μF capacitor
1 0.47-μF capacitor
1 1-μF capacitor
1 10-μF electrolytic or tantalum capacitor
1 150-Ω resistor
2 4.7-kΩ resistors
1 6.8-kΩ resistor
1 10-kΩ resistor
1 47-kΩ resistor
1 1kΩ pot

INTRODUCTION

Optimum amplitude modulation occurs when the peak-to-peak value of the modulating signal is equal to the peak value of the carrier. In that way, the modulating signal causes the carrier amplitude to vary from twice its unmodulated value to zero, as Fig. 2-1 illustrates. That is called a modulation index m of 1, or 100 percent. Of course, a modulation index of 0, or 0 percent modulation, occurs when the modulating signal amplitude is 0 and a constant amplitude carrier signal results. Intermediate values of modulation index or percent can be determined by mea-

4

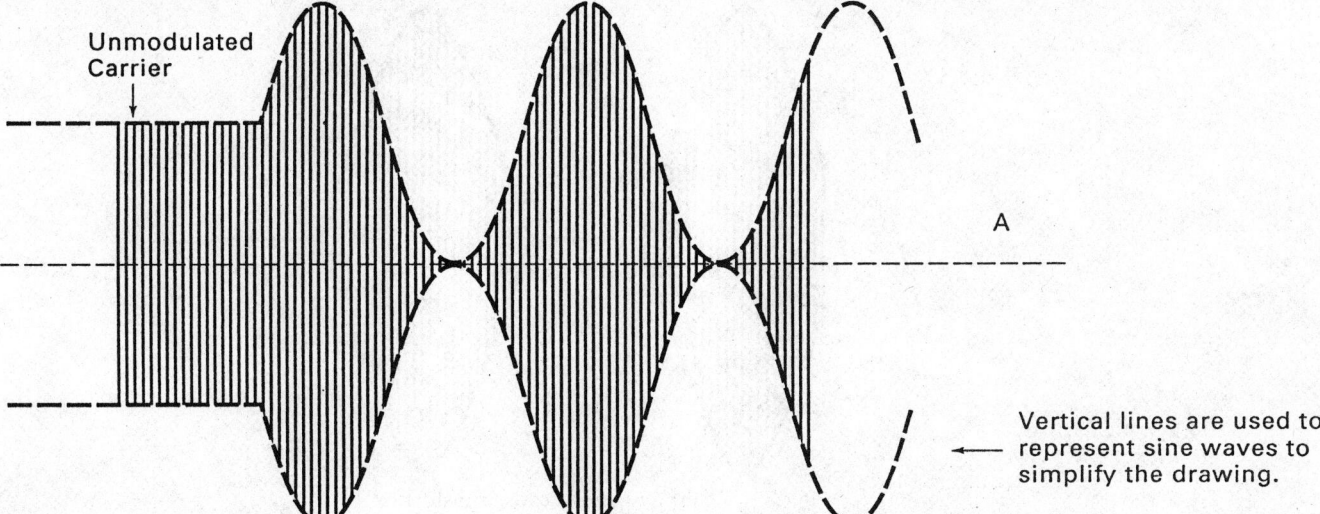

Unmodulated
Carrier

A

Vertical lines are used to
represent sine waves to
simplify the drawing.

Fig. 2-1 100 percent modulation.

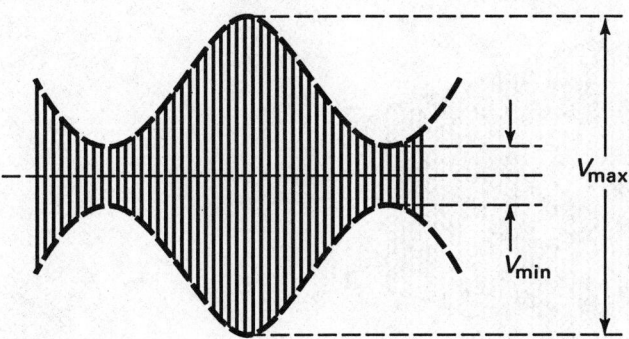

V_{max}

V_{min}

Fig. 2-2 Measuring voltages to calculate modulation index.

suring the amplitudes of the modulated signal, as shown in Fig. 2-2. The following formulas can then be used to compute the percent of modulation.

$$\text{Modulation index } m = \frac{V_{\text{max}} - V_{\text{min}}}{V_{\text{max}} + V_{\text{min}}}$$

$$\text{Percent modulation} = m \times 100$$

The most important criterion is to attempt to achieve 100 percent modulation but not exceed it. If the modulating signal amplitude is greater than the carrier amplitude, overmodulation will occur and the information signal will be distorted. Note how the peaks in Fig. 2-3 on the next page are clipped.

Another way to measure the modulation index or percent of modulation is to use a trapezoidal pattern that can be displayed on an oscilloscope. Such a pattern is shown in Fig. 2-4. To create it, the AM signal is connected to the vertical input as usual. The modulating information signal is connected to the external horizontal input and is used as the sweep signal. The scope is put into the external sweep or x vs. y display mode, and the vertical and horizontal gains are adjusted until the pattern appears. The V_{max} and V_{min} values can then be read from the pattern and used in the previously given formula to calculate the modulation index and percent of modulation.

The trapezoidal pattern method is usually preferred because mea-

5

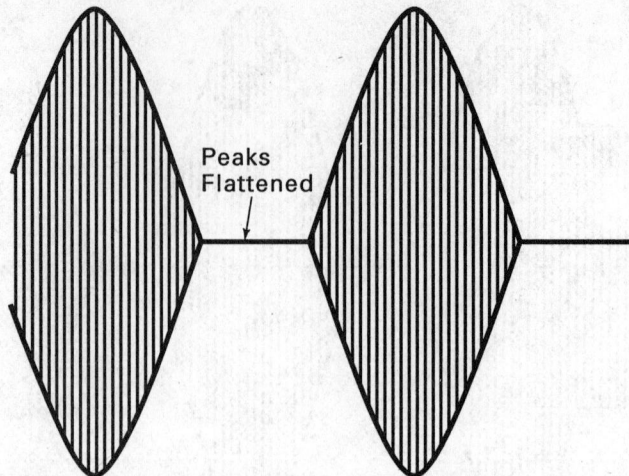

Fig. 2-3 Overmodulation caused distortion.

Peaks Flattened

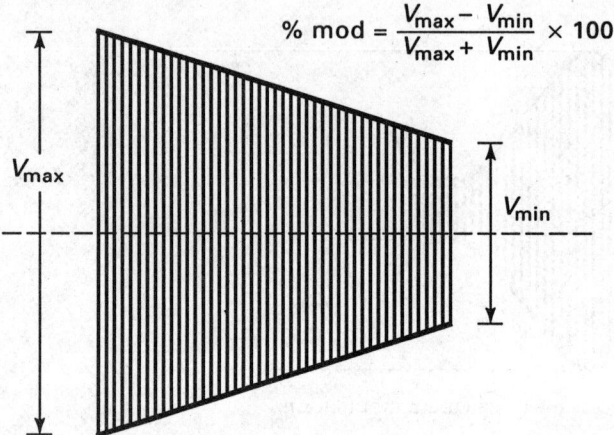

$$\% \text{ mod} = \frac{V_{max} - V_{min}}{V_{max} + V_{min}} \times 100$$

V_{max}

V_{min}

Fig. 2-4 Trapezoid pattern for measuring the percent of modulation.

surement of V_{max} and V_{min} is easier. Further, the slanted sides of the trapezoid readily show distortion. Straight sides mean no distortion; nonlinear sides indicate overmodulation or other distortion.

To demonstrate AM, you will use the XR-2206 Exar function generator IC. This chip contains an oscillator that produces sine, square, triangle, and sawtooth waves. The frequency is set by an external resistor and capacitor. The oscillator can also be voltage-controlled which permits its frequency to be varied. That allows sweep frequency or frequency modulation to be produced, but the function will not be used here.

The chip also contains an amplitude modulator circuit. The internal oscillator serves as the carrier, which can be amplitude-modulated by an external input signal. The modulator produces excellent AM, but the modulating signal amplitude required is higher than the theoretical value necessary to produce a given degree of modulation.

PROCEDURE

1. Construct the circuit shown in Fig. 2-5. You will use the external signal function generator with a 1-kHz output to modulate the car-

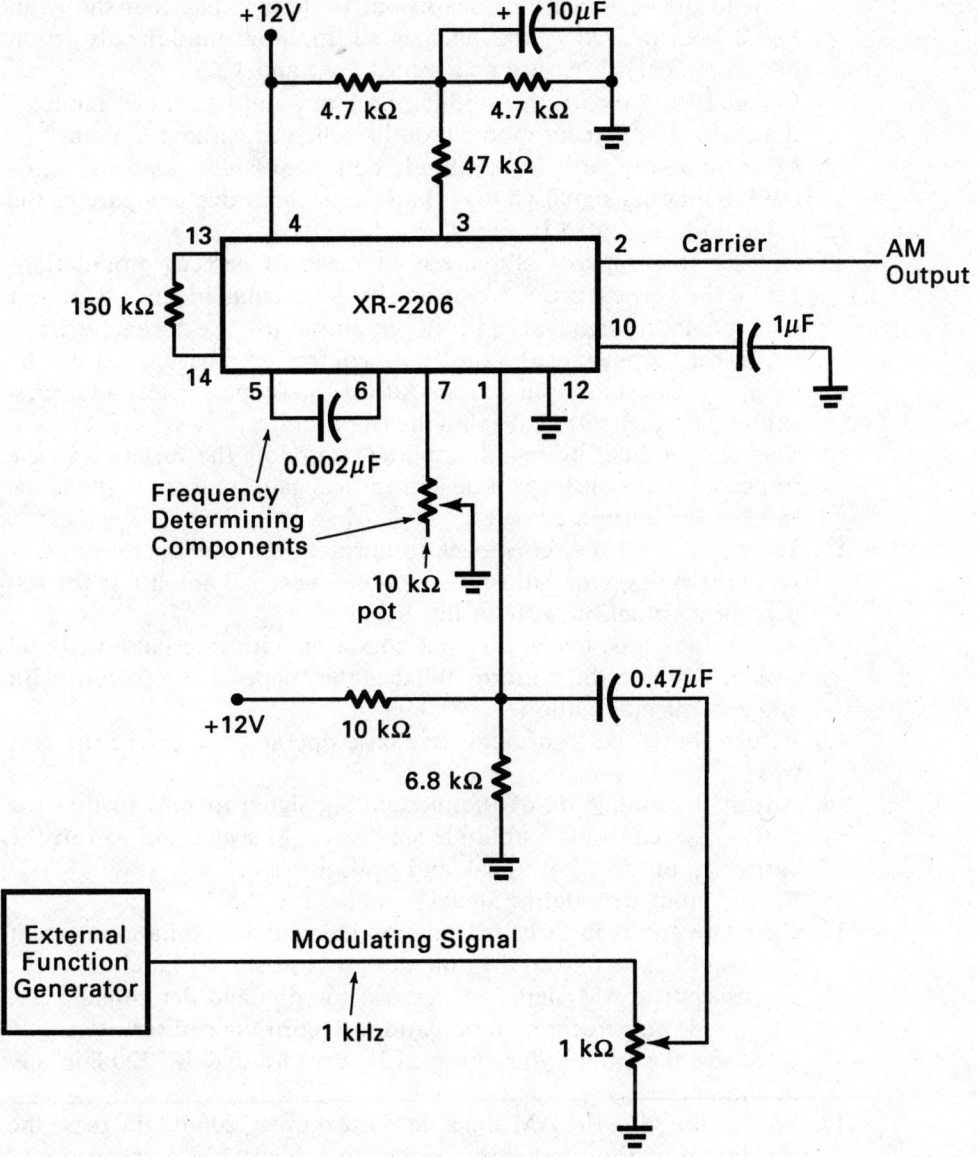

Fig. 2-5 Amplitude modulation of the 2206 function generator IC.

rier produced by the 2206 IC function generator. The 2206 chip contains a built-in amplitude modulator.

2. Apply power to the circuit. Adjust the dc supply voltage to 12 V. Observe the waveform at the output at pin 2 of the carrier function generator IC. Reduce the modulating signal input from the external function generator to zero with the 1-kΩ pot or the amplitude control on the function generator. While measuring the output at pin 2 with a frequency counter, adjust the 100-kΩ pot for a frequency of 40 kHz. (Note: If no counter is available, use the calibrated horizontal sweep of the scope to set the frequency.)

3. Measure and record the carrier output voltage, peak and peak-to-peak, with the oscilloscope.

4. Use the value obtained in step 3 to calculate and record the amount of voltage required to 100 percent modulate this carrier.

5. Vary the amplitude of the modulating signal from the external function generator by using the 1-kΩ pot and/or the external amplitude control from zero to about 5 $V_{\text{p-p}}$. Note the effect on the AM signal.

6. Adjust the modulating signal amplitude for maximum value just

prior to the occurrence of distortion. If clipping like that shown in Fig. 2-3 occurs, reduce the input signal amplitude until the distortion just disappears. Measure and record V_{max} and V_{min}.

7. Calculate and record the modulation index and percent of modulation. Can 100 percent modulation be achieved without clipping?

8. Measure and record the amplitude both peak and peak-to-peak, of the modulating signal voltage. How does this value compare to the value you calculated in step 4?

9. Display the trapezoidal pattern to measure percent modulation. Leave the vertical scope input on the AM signal at pin 2. Connect the modulating signal at the 1-kΩ pot output to the external horizontal (sweep) input. Set the time base control or other switch on the scope for external input sweep. Adjust the scope vertical and horizontal gain controls to display the trapezoid.

10. Vary the modulating signal amplitude and note the variation in the trapezoid. Vary the horizontal and vertical gain controls on the scope to keep the pattern centered.

11. Temporarily set the scope back to normal time base operation. Observe the AM wave. Adjust the modulating signal amplitude for 100 percent modulation without distortion.

12. Return the scope to the previous condition with external horizontal sweep. Observe the pattern. What is the shape of the pattern with 100 percent modulation?

13. Return the scope to normal time base operation. Observe the AM wave.

14. Adjust the amplitude of the modulating signal to zero so that the output is a constant amplitude sine wave. Measure and record the carrier amplitude, both peak and peak-to-peak.

15. Set the input modulating signal amplitude to 0.5 V_{p-p}.

16. Calculate and record the modulation index and modulation percent you would expect from the voltages given in steps 14 and 15.

17. Now view the AM signal on the oscilloscope and determine V_{max}, V_{min}, and the percent of modulation. Record your observations.

18. Compare the values you obtained in steps 16 and 18. Explain any discrepancies.

19. While observing the AM signal at pin 2 on the 2206 IC, increase the amplitude of the modulating signal to achieve 100 percent modulation. Then continue increasing the modulating signal amplitude until distortion occurs. Describe the effect you observe when overmodulation occurs.

20. Do not disassemble the modulator circuit, because you will use it in future experiments. For now, turn off the power.

REVIEW QUESTIONS

Read the question and write the answer on a separate sheet of paper.

1. On an AM wave, V_{max} = 2 V and V_{min} = 0.5 V. The modulation index is _____.
   a. 0.25   c. 0.6
   b. 0.4   d. 0.85

2. For 100 percent modulation, what is the relationship between the carrier voltage V_c and the modulating signal voltage V_m? _____
   a. $V_m = V_c = 0$ c. $V_m > V_c$
   b. $V_m = V_c$ d. $V_m < V_c$

3. When overmodulation occurs, _____.
   a. $V_m = 2V_c$
   b. $V_m = V_c$
   c. $V_m < V_c$
   d. $V_m > V_c$

4. When you are using a trapezoidal pattern to measure modulation percentage, 100 percent modulation is indicated by what shape of pattern? _____
   a. Triangular
   b. Rhomboid
   c. Square
   d. Rectangle

5. (True or False) The frequency of the modulating affects the percent of modulation. _____

ACTIVITY 2-3 (OPTIONAL)
LAB EXPERIMENT:
SPECTRUM ANALYZER

PURPOSE

To demonstrate the operation and application of a spectrum analyzer.

MATERIALS

All the equipment and circuitry used in Activity 2-2 plus a spectrum analyzer capable of measuring and displaying low-frequency signals.

INTRODUCTION

A spectrum analyzer is a test instrument designed to measure and display signals in a frequency domain format. Like an oscilloscope, the spectrum analyzer uses a CRT display, but, unlike a scope, it displays signal amplitude with respect to frequency rather than time. Complex waveforms made up of many signals are shown as vertical lines whose heights indicate signal strength and whose horizontal positions indicate frequency. In other words, the spectrum analyzer effectively displays a complex signal as its individual sinusoidal components.

The spectrum analyzer is an ideal test instrument for use with modulated signals because it allows you to see the carrier and sidebands resulting from the modulation. It makes measurement and troubleshooting faster and easier.

In this experiment, you will become familiar with spectrum analyzer operation and use. You will display the spectrum produced by the 2206 amplitude-modulated function generator you analyzed in the preceding experiment.

PROCEDURE

1. Acquire the operation manual for the spectrum analyzer you will be using. Take some time to read and study it. Concentrate on the following:
   a. How does the spectrum analyzer work? Analyze its operation in terms of a block diagram of the major circuits.

b. How do you use the spectrum analyzer? What are the various controls, inputs, and output? What are the procedures for setting up, calibrating, and making measurements with the unit?

c. What are the unit's specifications?

2. Set up the spectrum and become familiar with the controls and how to display a signal.

3. Apply power to the 2206 function generator IC. Adjust the output for a frequency of 35 kHz. Set the modulating signal frequency to 1 kHz. While observing the output at pin 2 of the 2206 (see Fig. 2-5), increase the modulating signal amplitude until 100 percent modulation is achieved. Predict what signal the spectrum analyzer will display and sketch it.

4. Apply the AM signal to the spectrum analyzer, and adjust the controls to display the signal. How does it compare to the prediction you made?

5. Reduce the modulating signal amplitude and note the effect it has on sideband amplitude.

6. While observing the AM signal, increase the modulating signal amplitude until overmodulation occurs. The signal should be as clipped and distorted as possible. Observe the spectrum analyzer display. Explain how the distortion is indicated.

REVIEW QUESTIONS

Read the question and write the answer on a separate sheet of paper.

1. (True or False) The spectrum analyzer displays the amplitudes of the individual sinusoidal components of a complex signal as determined by the Fourier theory. _____

2. With 60 percent modulation and a 500-mV carrier, the sideband amplitude for a sine wave modulating signal is _____.
   a. 150 mV
   b. 250 mV
   c. 300 mV
   d. 500 mV

3. The largest component in an AM signal as displayed on a spectrum analyzer is the _____.
   a. Upper sideband
   b. Lower sideband
   c. Carrier
   d. Modulating signal

4. Distortion from overmodulation is displayed by a spectrum analyzer as _____.
   a. Greater carrier power
   b. Higher sideband amplitudes
   c. Lower carrier and sideband amplitudes
   d. Additional harmonic-related sidebands

5. The horizontal sweep calibration of a spectrum analyzer is given in units of _____.
   a. Time
   b. Frequency
   c. Voltage
   d. Power

CHAPTER | 3

Amplitude Modulator Circuits

**ACTIVITY 3-1
TEST: AMPLITUDE
MODULATOR CIRCUITS**

Read the question and write the answer on a separate sheet of paper.

1. The nonlinear component most often used as an amplitude modulator or mixer is the _____.

2. The pin diode is used as a variable _____.

3. Linear power amplifiers are used in _____.
   a. High-level modulation
   b. Low-level modulation

4. The final power amplifier in a high-level AM system usually operates _____.
   a. Class A
   b. Class B
   c. Class C

5. The simplest AM receiver is a(n) _____.

6. The circuit at the output of an AM diode detector is a(n) _____.

7. The output of a balanced modulator is _____.
   a. AM
   b. DSB
   c. SSB

8. In a lattice modulator, the diodes act as _____.

9. The basic circuit in the 1496 IC balanced modulator is a(n) _____.

10. The most common way of generating SSB is by _____.

11. Name three kinds of filters used in SSB transmitters. _____

12. Name the method of SSB generation that does not use filtering. _____

13. (True or False) Linear mixing and frequency conversion are the same.

14. A popular mixer for microwaves is the _____.

15. A converter combines a(n) _____ and _____ in a single stage.

ACTIVITY 3-2
LAB EXPERIMENT:
DIODE MODULATOR AND MIXER

PURPOSE

To demonstrate how a diode can be used to perform amplitude modulation mixing.

MATERIALS

You will use all of the components and test equipment used in Activity 2-2 plus the following:

Qty.

1	741 IC op amp
1	1N4149 or 1N4148 silicon diode
1	2-mH inductor
3	0.1-μF capacitors
1	0.01-μF capacitor
4	10-kΩ resistors
1	15-kΩ resistor

INTRODUCTION

Amplitude modulation and mixing are actually one and the same process. Two or more signals are combined in a circuit that produces new signals at different frequencies. The amplitude modulator generates sidebands whose frequencies are the sum and difference of the frequencies of the carrier of the modulating signal. A mixer produces similar signals. The output of a mixer consists of the two original input signals plus the sum and difference frequencies, which are, of course, the same as the sidebands.

One of the simplest ways to perform mixing or amplitude modulation is to linearly combine the two signals in a resistive network or op amp summer and then feed them to a diode rectifier. The output of the rectifier consists of a tuned circuit or filter which selects the desired output frequency. In this experiment, you will demonstrate the use of a diode for amplitude modulation and mixing.

PROCEDURE

Modulator

1. Modify the function generator circuit you used previously in Activity 2-2; the revised circuit is shown in Fig. 3-1. First remove the 10- and 6.8-kΩ resistors connected to pin 1 of the 2206. Also remove the 0.47-μF capacitor and the 1-kΩ pot. Then connect pin 1 of the 2206 to ground. This modification removes the amplitude modulation capability of the 2206. In this experiment you will use the function generator simply as a carrier oscillator.
2. Construct the remaining portion of the circuit shown in Fig. 3-1. The circuit consists of an op amp summer used as a linear mixer and a diode modulator mixer circuit. This first circuit will be used for amplitude modulation.
3. Apply power to the circuit. Monitor the output of the 2206 function generator at pin 2. You should observe a sine wave output with an

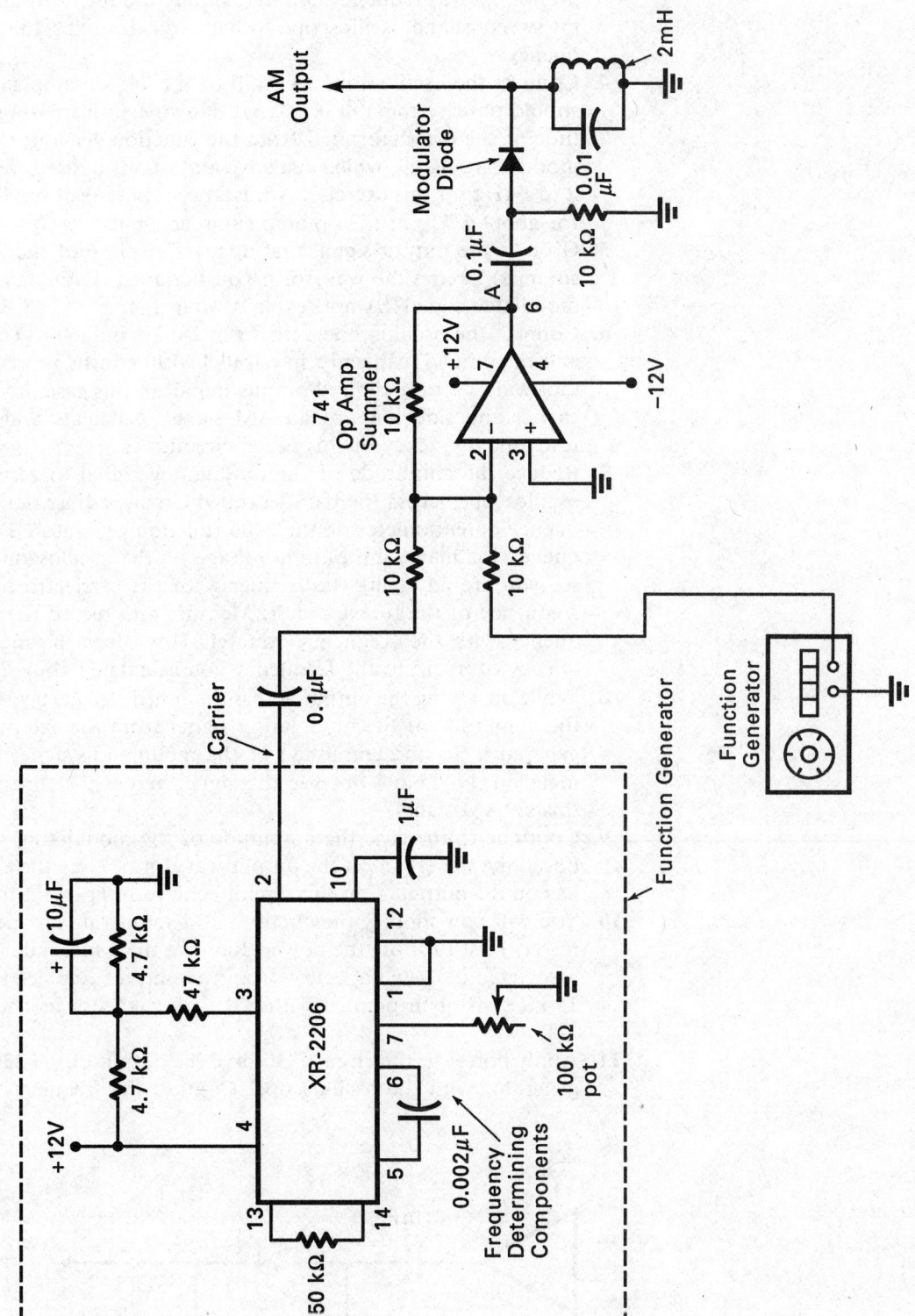

Fig. 3-1 Diode (AM) modulator circuit.

13

amplitude of approximately 4 $V_{\text{p-p}}$. The input signal from the external function generator should be zero at this time. Next, connect the frequency counter to the function generator output. Then adjust the 100-kΩ potentiometer for an output frequency of 40 kHz. If you do not have a frequency counter, simply use the calibrated horizontal sweep of the oscilloscope to help you in setting the correct frequency.

4. Connect the oscilloscope to pin 6 of the 741 op amp summer. You should observe the 40-kHz signal. Now begin increasing the amplitude of the external signal from the function generator. This is the modulating signal, which can have almost any desired frequency; set it to 1 kHz for this exercise. Adjust the amplitude of the 1-kHz signal for about 5 $V_{\text{p-p}}$ at the op amp summer input.

5. Observe the output signal that appears at pin 6 of the 741 op amp summer. Sketch the waveform you obtained. Is this an amplitude-modulated signal? If not, explain what it is.

6. Connect the oscilloscope across the 2-mHz inductor. This inductor and the 0.01-µF capacitor in parallel with it form a resonant circuit that will select the desired output signal. In this case, it will pass the carrier and sidebands of an AM signal. Calculate and record the resonant frequency of this tuned circuit.

7. Reduce the amplitude of the modulating signal to zero. With the oscilloscope across the parallel tuned circuit, adjust the 100-kΩ frequency potentiometer on the 2206 function generator. Tune the frequency for maximum output voltage on the oscilloscope. In doing so, you are adjusting the frequency of the carrier to the resonant frequency of the tuned circuit. Measure and record the output frequency with the frequency counter. How does the measured frequency compare to the frequency you calculated above?

8. While observing the output across the tuned circuit, begin increasing the amplitude of the modulating signal from the external function generator. Set the amplitude of the modulating signal to approximately 1 $V_{\text{p-p}}$. Now observe the signal across the tuned circuit. Is this an AM signal?

9. Continue to increase the amplitude of the modulating signal while observing the effect on the output waveform. Note the effect that is has on the output. Can this circuit achieve 100 percent modulation?

10. You will now modify the circuit and demonstrate its operation as a mixer. First turn off the power. Remove the 2-mH inductor and the associated 0.01-µF capacitor from the output. Replace them with a 15-kΩ resistor in parallel with a 0.1-µF capacitor as shown in Fig. 3-2.

11. Apply power to the circuit. Observe the output of the 2206 function generator with the oscilloscope. Connect the frequency counter to

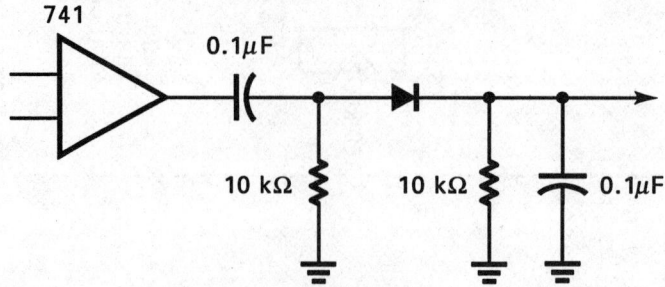

Fig. 3-2 Modifying the diode circuit for mixer operation.

14

the output of the 2206. Adjust the 100-kΩ pot on the 2206 for a frequency of 20 kHz. This is one input to the mixer.

12. Next, observe the output from the external function generator at the second input to the op amp summer. With the frequency counter, adjust the external function generator output for a frequency of 21 kHz. Set the output amplitude of the external function generator to a value of 1 V_{p-p}. This is the second input to the mixer.

13. Based on what you have learned about mixers, calculate and record the output frequencies that would occur at the diode mixer output.

14. Considering the components connected to the output of the diode mixer, which of the output signals you predicted in the preceding step would appear at the output?

15. Connect the oscilloscope and the frequency counter to the 15-kΩ load. Observe the signal at the output of the mixer. What are its shape and frequency?

REVIEW QUESTIONS

Read the question and write the answer on a separate sheet of paper.

1. What mathematical process is involved in linear mixing? _____
   a. Addition
   b. Subtraction
   c. Multiplication
   d. Division

2. What mathematical process is involved in amplitude modulation or nonlinear mixing? _____
   a. Addition
   b. Subtraction
   c. Multiplication
   d. Division

3. (True of False) Nonlinear mixing and AM are the same process. _____

4. What is the function of the 0.1-μF capacitor shown in Fig. 3-2? _____
   a. Resonant circuit   c. High-pass filter
   b. Noise elimination   d. Low-pass filter

5. If the 0.1-μF output capacitor in Fig. 3-2 is replaced with the circuit shown in Fig. 3-3, what output signal will you expect to see? _____
   a. Difference   c. Carrier
   b. Sum   d. Modulating signal

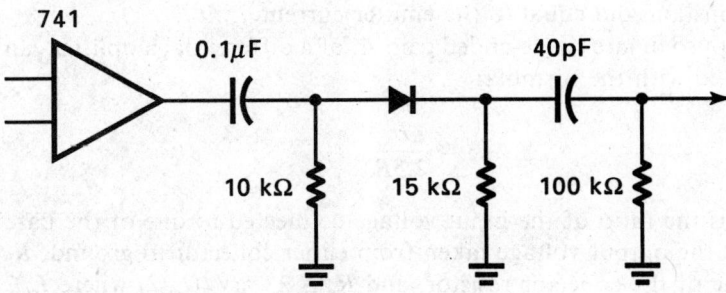

Fig. 3-3 Circuit for question 5.

ACTIVITY 3-3
LAB EXPERIMENT:
DIFFERENTIAL AMPLIFIER
MODULATOR

PURPOSE

To demonstrate the use of a differential amplifier as an amplitude modulator.

MATERIALS

You will need the 2206 function generator circuit and all of the other components and test equipment listed for Activity 2-2. In addition, you will need:

Qty.

3   NPN transistors (2N4401, 2N4124, 2N3904, MPSA20, etc.)
1   741 op amp
1   2-mH inductor
3   0.1-μF capacitors
1   0.47-μF capacitor
1   4.7-kΩ resistor
1   6.8-kΩ resistor
2   15-kΩ resistors
1   47-kΩ resistor
1   10-kΩ pot

INTRODUCTION

A differential amplifier makes an ideal amplitude modulator for low-level circuits, and a standard circuit of one is shown in Fig. 3-4. The differential transistors Q_1 and Q_2 are connected so their emitters are common. Inputs are applied to the base of one transistor or the bases of both transistors. Two outputs at the transistor collectors are provided. A load can be connected between the two collectors to provide a differential output. Loads can also be connected between the two collectors for balanced operation or between either collector and ground to provide single-ended operation.

The common current I_e to the differential transistors is supplied by a constant-current source, transistor Q_3. Bias resistors R_1 through R_3 and the negative supply voltage V_{EE} set the current value. The emitter current I_e remains constant despite any input circuit variations. The current splits between the two differential transistors. As the base inputs to Q_1 or Q_2 vary, the emitter current divides between the two transistors in proportion to the signal applied. The sum of the currents through Q_1 and Q_2 remains constant and equal to the emitter current.

The approximate single-ended gain A_s of a differential amplifier can be computed with the formula:

$$A_s = \frac{R_c}{2.5R_e}$$

This gain is the ratio of the input voltage connected to one of the base inputs and the output voltage taken from either collector to ground. R_c is the value of the collector resistor, and R_e is 25 mV/(I_e/2) where I_e/2 is the current through each differential transistor.

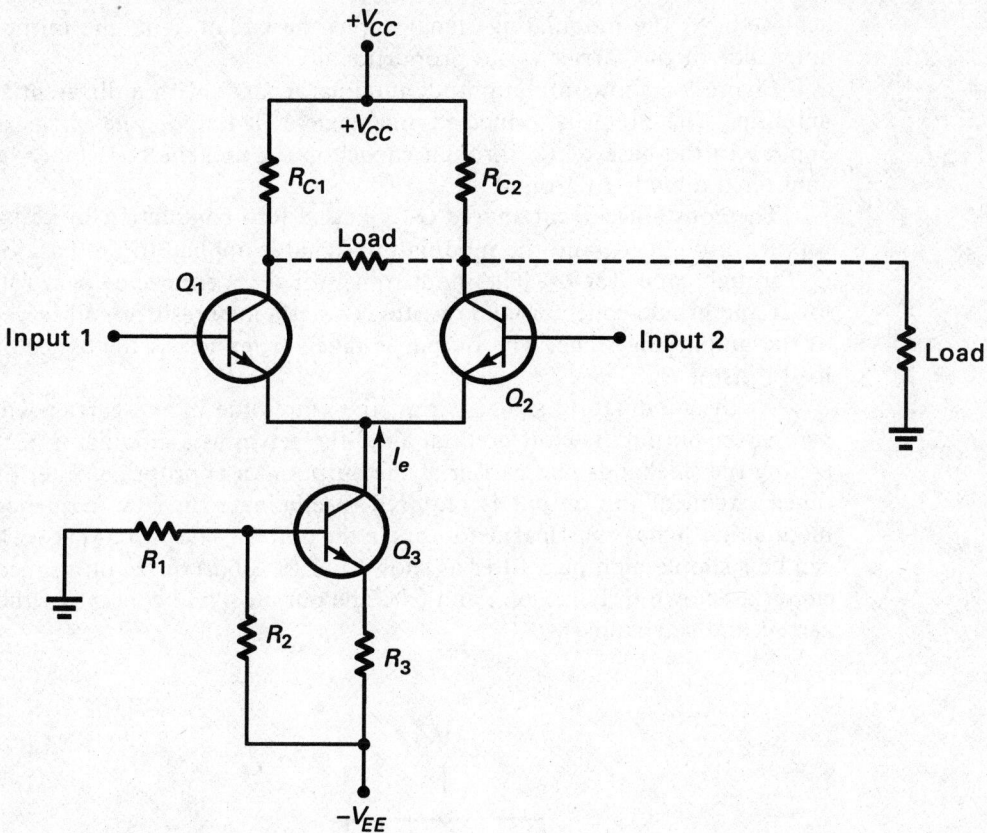

Fig. 3-4 Differential amplifier.

For example, if $R_c = 10$ kΩ and $I_e/2 = 1.2$ mA,

$$R_e = \frac{25 \times 10^{-3}}{1.2 \times 10^{-3}} = 20.83$$

Therefore, the gain is:

$$A_s = \frac{R_c}{2.5R_e} = \frac{10,000}{2.5(20.83)}$$

$$A_s = \frac{10,000}{52.1} = 192$$

When the gain formula is rearranged algebraically in terms of I_e, it becomes:

$$A_s = \frac{R_c I_e}{125}$$

where I_e is the current from the current source, in milliamperes.

The circuit gain is directly proportional to the emitter current and collector resistance. Under ordinary conditions, both I_e and R_c are fixed values, thereby providing a fixed circuit gain. In amplitude modulation, however, modifications are made to the circuit to cause the emitter current to vary. That is done by applying the modulating or information signal to the constant-current source Q_3, which causes the emitter current to vary linearly and thereby change the emitter current. Changing the emitter current, in turn, changes the circuit gain. By using the differential amplifier to amplify a higher-frequency carrier signal, modulation is achieved. As the modulating signal varies the circuit gain, the output amplitude of the carrier varies proportionally.

Figure 3-5 shows an amplitude modulator made with a differential amplifier. The circuit is connected in a single-ended mode; the carrier is applied to the base of Q_1 through capacitor C_1; and the base of Q_2 is connected directly to ground.

The constant-current source Q_3 is biased into conduction by resistors R_1 through R_3, and the modulating signal is applied to the base of Q_3 through capacitor C_2. The input transistor Q_1 is connected as a follower and has no collector load resistor. Q_2 has a load resistor and serves as the amplifying stage. The output is taken from across the collector load resistor R_5.

As the modulating signal varies, the amplitude of the carrier will vary at the output. The differential amplifier serves as a modulator generating the sidebands and carrier at the output. An appropriate filter or tuned circuit at the output is required to eliminate the low-frequency modulating signal, which also appears in the output. The output network can be a simple high-pass filter as shown in Fig. 3-6(a) or (b) or a tuned circuit as shown in Fig. 3-6(c) and (d). The output signal consists of the carrier and sidebands.

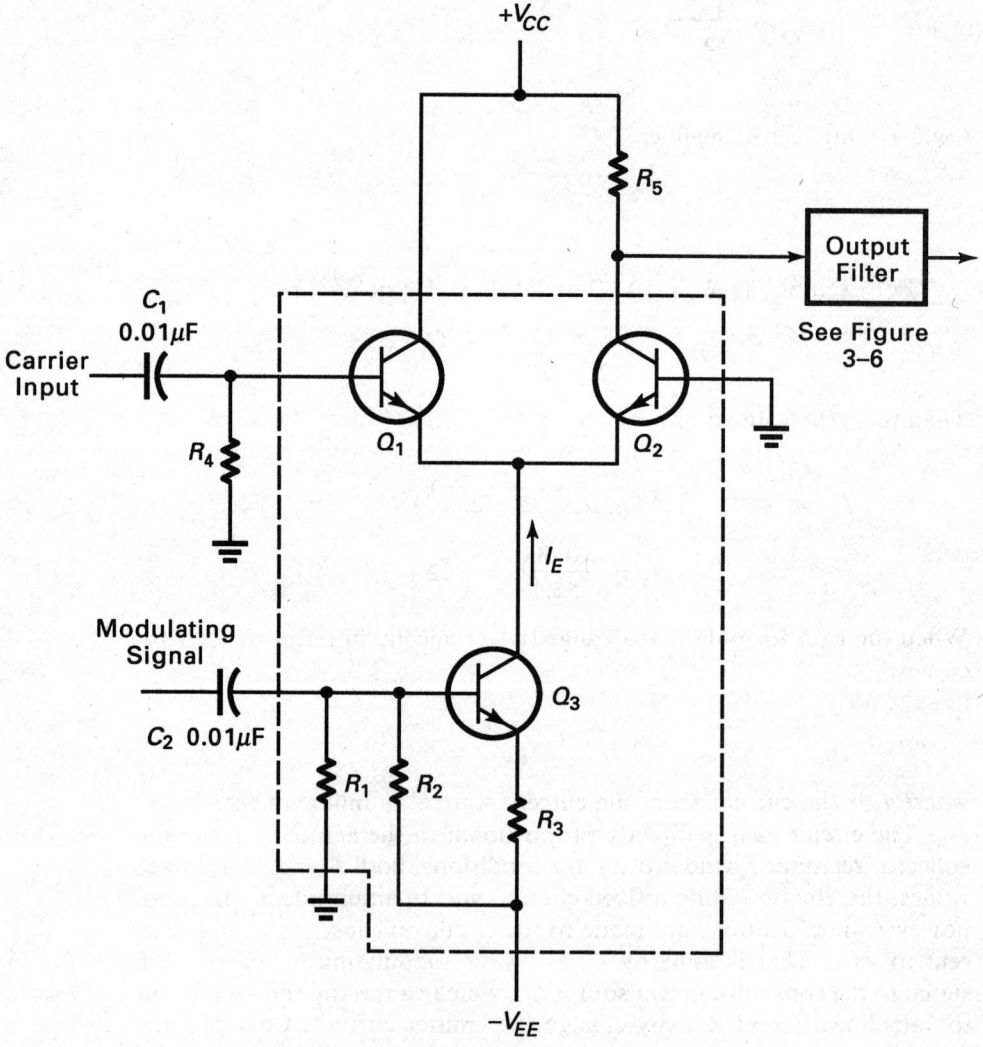

Fig. 3-5 Differential amplifier amplitude modulator.

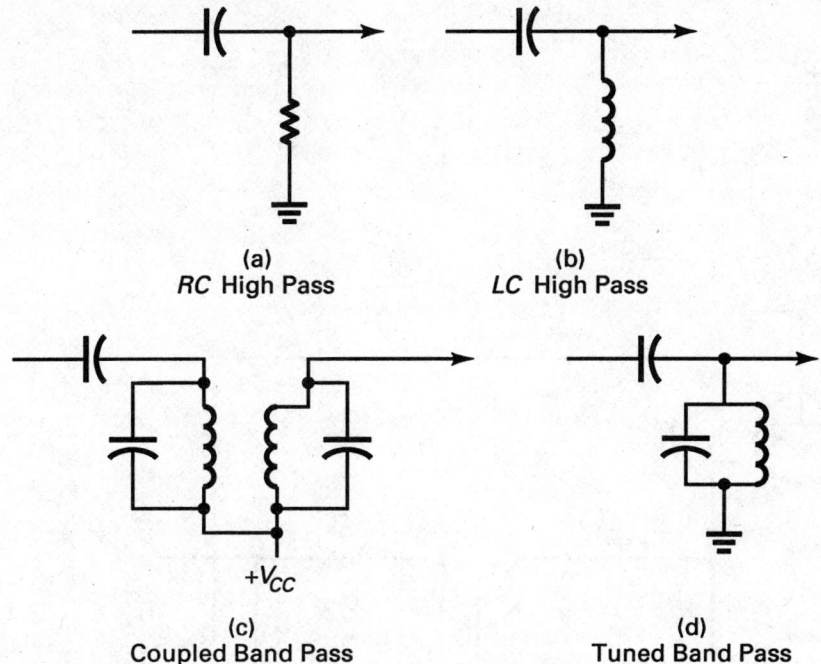

(a)
RC High Pass

(b)
LC High Pass

(c)
Coupled Band Pass

(d)
Tuned Band Pass

Fig. 3-6 Output filters for differential amplifier amplitude modulator.

It is the amplitude of the modulating signal that determines the percent of modulation, and the signal is increased until 100 percent modulation is obtained. The negative-going portion of the modulating signal is capable of reducing the gain of the differential amplifier to zero; the positive-going portions can double the circuit gain. Therefore, the circuit is capable of producing 100 percent modulation. Excessive modulating signal input will cause clipping and distortion as in other amplitude modulators.

In this experiment, you will demonstrate the use of a differential amplifier as an amplitude modulator. Such a circuit is widely used; for example, one version of it produces AM in the 2206 IC function generator you used in the preceding experiment.

PROCEDURE

1. Wire the differential amplifier amplitude modulator circuit shown in Fig. 3-7. Q_1–Q_3 can be any common NPN transistor. The carrier is supplied by the 2206 function generator, and the modulating signal is supplied by an external function generator. Initially, do not connect the 0.1-μF capacitor and the tuned circuit to the collector of Q_3.

2. Reduce the output of the external function generator supplying the modulating signal to zero. With an oscilloscope, check at the base of Q_3 to ensure that no ac signal is present.

3. Observe the carrier at the output of the 741 op amp follower. Using the 100-kΩ pot, adjust the frequency of the 2206 to 12 kHz. Apply approximately 100 mV of carrier signal to the modulator at the base of Q_1 by adjusting pot R_4. Observe the differential amplifier output at the collector of Q_2. It should be an undistorted sine wave; if it is not, reduce the input to Q_1 with R_4 until any distortion is eliminated.

4. Connect the 0.1-μF capacitor and tuned circuit to the collector of Q_3 as Fig. 3-7 shows.

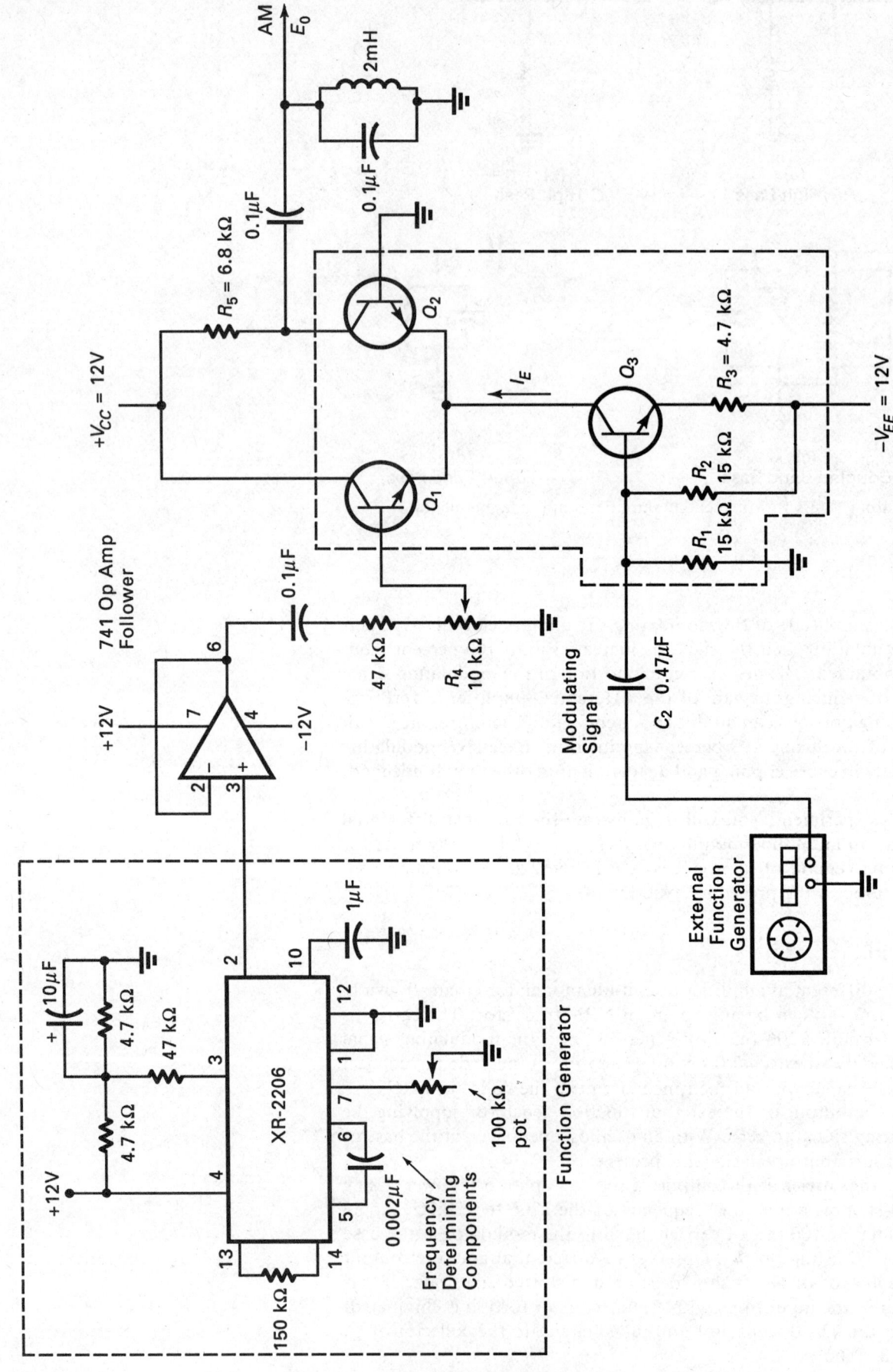

Fig. 3-7 AM with a differential amplifier.

20

5. Now connect the oscilloscope across the tuned output circuit. Vary the 100-kΩ pot on the 2206 function generator to set the carrier frequency to the resonant frequency. Tune for maximum output voltage.

6. Given the values in Fig. 3-7, calculate the resonant frequency f_r. Measure the carrier frequency with a frequency counter.

7. Set the modulating signal frequency to 300 Hz. While observing the carrier output across the tuned circuit, begin increasing the amplitude of the modulating signal from the external function generator. Note the effect on the output signal. You should see a clean amplitude-modulated wave.

8. Adjust the output waveform for 100 percent modulation.

9. Continue to increase the amplitude of the modulating signal to determine if it is possible to overmodulate the differential amplifier and produce distortion and clipping. Is it?

REVIEW QUESTIONS

Read the question and write the answer on a separate sheet of paper.

1. In a differential amplifier, AM is achieved by varying what characteristic of the circuit with the modulating signal? _____
   a. Bandwidth
   b. Gain
   c. Resonant frequency
   d. Distortion

2. The modulating signal causes what parameter in the differential amplifier to vary? _____
   a. Base voltage of input transistor
   b. Emitter resistance of current source
   c. Collector voltage of Q_1
   d. Emitter current of the differential pair

3. The output tuned circuit is needed to _____.
   a. Increase gain
   b. Suppress the carrier
   c. Eliminate the modulating signal
   d. Eliminate the sidebands

4. (True or False) 100 percent modulation can be achieved with the differential amplifier modulator. _____

5. Assuming I_e is constant, what component in Fig. 3-7 would you change to increase the gain of the circuit? _____
   a. R_5
   b. R_2
   c. R_4
   d. R_3

ACTIVITY 3-4
LAB EXPERIMENT:
DIODE DETECTOR

PURPOSE

To construct a diode detector circuit and demonstrate its use in demodulating an AM signal.

MATERIALS

In addition to the equipment and components you used in Activity 2-2, you will need the following:

Qty.

1   741 IC op amp
1   Silicon signal diode (1N914, 1N4148, 1N4149, etc.)
1   470-pF capacitor
1   0.005-μF capacitor
1   0.01-μF capacitor
1   1-kΩ resistor
1   10-kΩ resistor
2   100-kΩ resistors
1   10-kΩ pot

INTRODUCTION

The process of recovering the original information signal from an AM waveform is known as demodulation. A variety of circuits can be used to demodulate an AM signal, but the one that is simplest and most widely used is the diode detector. A typical circuit is shown in Fig. 3-8. The rectifier diode D_1 removes one-half of the AM signal, and the capacitor

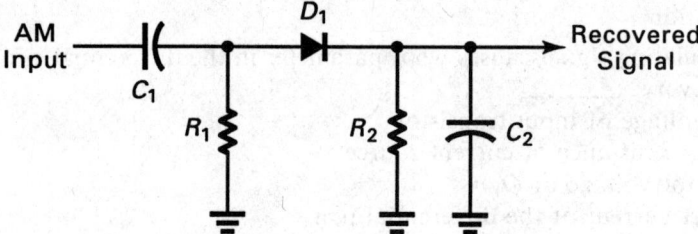

Fig. 3-8   Diode AM detector circuit.

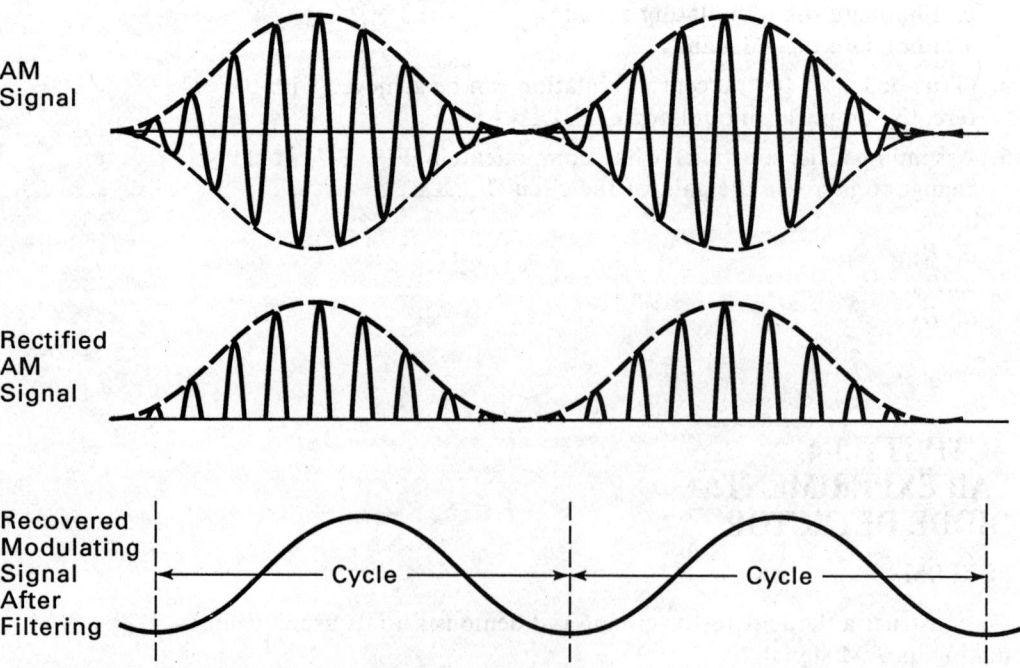

Fig. 3-9   Demodulation of an AM signal.

C_2 across the load R_2 acts as a low-pass filter to remove the carrier, leaving only the original modulating signal. The waveforms shown in Fig. 3-9 illustrate this process.

The value of capacitor C_2 is critical in that it determines the degree to which the carrier is filtered out. The capacitor must be large enough to filter the carrier out, but if it is too large, it will cause distortion of the recovered signal. Thus, a compromise value is selected.

In this experiment, you will construct a diode detector circuit, show how it demodulates an AM signal, and demonstrate the effect of different values of filter capacitance on the signal waveshape.

PROCEDURE

1. The modulator circuit you built in Activity 2-2 will be used as the AM signal source in this experiment. Rewire the circuit to conform to Fig. 3-10. The modulating signal will come from an external func-

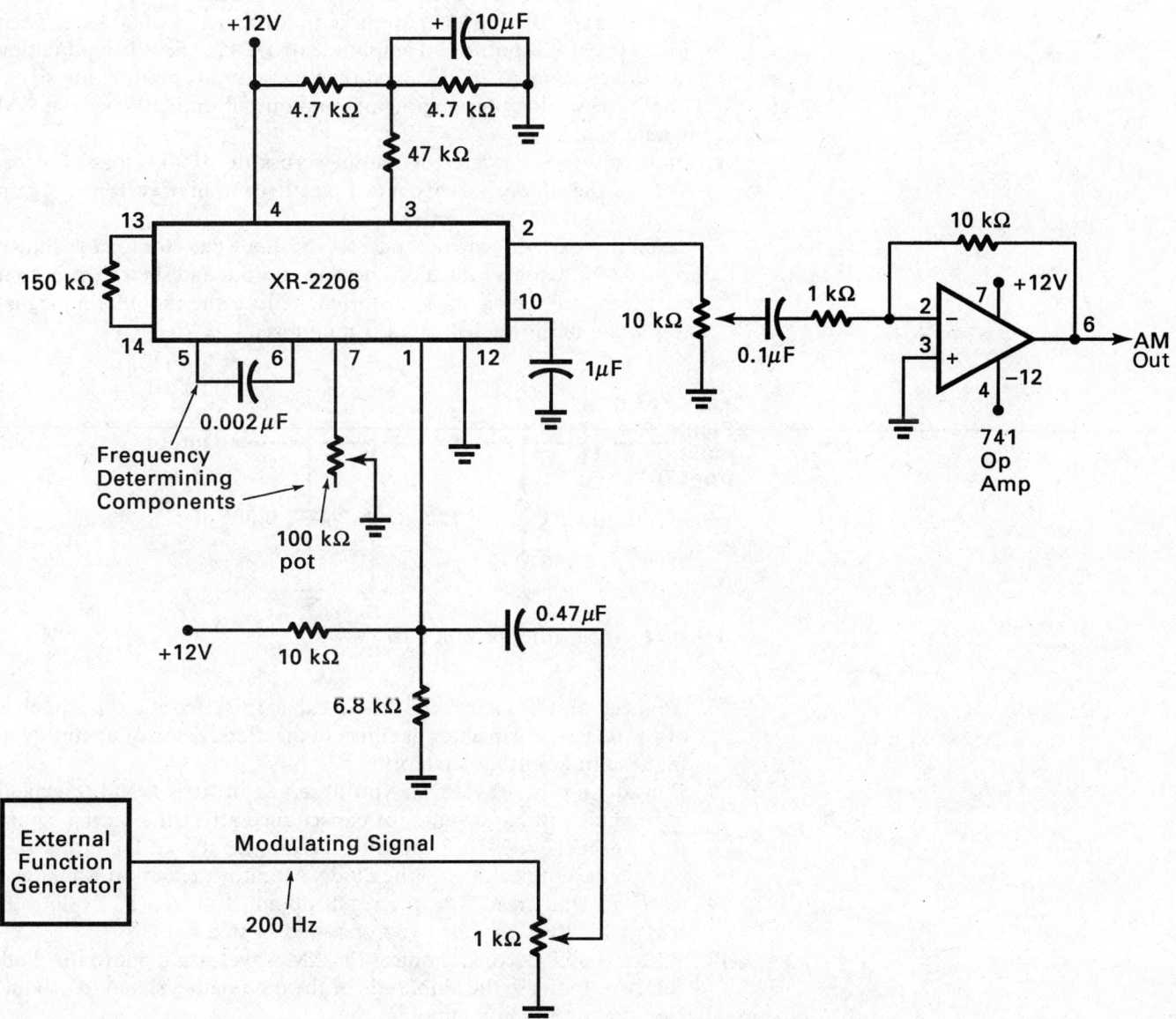

Fig. 3-10 Amplitude modulation of the 2206 function generator IC.

23

tion generator; the 741 op amp amplifies the AM signal to a level suitable for detection; and the 10-kΩ pot is used to adjust the carrier level.

2. Next, construct the diode detector circuit shown in Fig. 3-11. Note that a filter capacitor is not connected at this time.

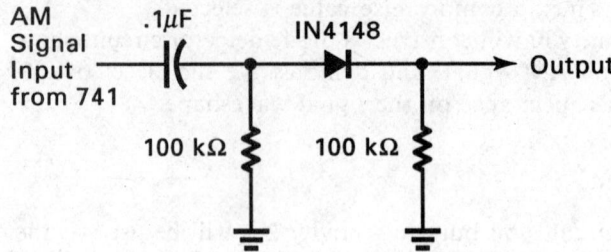

Fig. 3-11   Diode detector without filter capacitor.

3. Apply power to the circuit. Reduce the modulating signal amplitude to zero. While observing the carrier output of the 741 op amp, adjust the 100-kΩ pot on the 2206 for an output frequency of 30 kHz. Then adjust the 10-kΩ pot for an amplitude of 1.5 V_{p-p}. Set the modulating signal frequency to 200 Hz, and then increase the modulating signal amplitude while observing the op amp output until 100 percent AM is obtained.

4. Observe the diode detector output across the 100-kΩ load resistor. Reverse the diode polarity and note the output waveform. Again reverse the diode connections.

5. Connect a 0.005-μF capacitor across the diode detector load as shown in Fig. 3-12. Observe the diode detector output and sketch the output waveform. If the signal is distorted, reduce the modulating signal amplitude until the distortion is minimized.

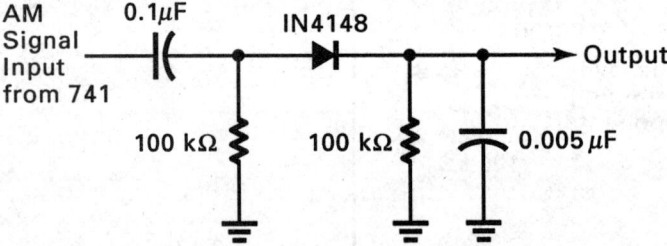

Fig. 3-12   Diode detector with filter.

6. Connect the 470-pF and the 0.1-μF capacitors, one at a time, in place of the 0.005-μF capacitor as filters in the diode detector circuit. Note the resulting output waveforms.

7. Considering the waveforms you observed in steps 5 and 6, explain how small and large values of capacitance affect the output signal.

8. Reconnect the 0.005-μF capacitor across the 100-kΩ load, and then reverse the connections to the diode. Monitor the output waveform.

9. Compare your results to the result obtained in step 5. Explain the reason for the waveform you observed in step 8.

10. With the oscilloscope, monitor the AM waveform input to the diode detector. Increase the amplitude of the modulating signal to produce overmodulation and clipping.

11. Observe the recovered waveform. Explain the results you obtained.

12. While observing the recovered signal at the output of the diode detector, vary the modulating signal amplitude from zero until distortion occurs. Explain what effect the percent of modulation has on the output signal.

REVIEW QUESTIONS

Read the question and write the answer on a separate sheet of paper.

1. If the filter capacitor in a diode detector is too small, the _____.
   a. Carrier is not removed
   b. Recovered signal is distorted
   c. Carrier is distorted
   d. Signal amplitude is reduced

2. If the filter capacitor in a diode detector is too large, the _____.
   a. Carrier is not filtered out
   b. Recovered signal is distorted
   c. Carrier is distorted
   d. Signal amplitude is reduced

3. Decreasing the percent of modulation causes the demodulated signal amplitude to _____.
   a. Increase
   b. Decrease
   c. Remain the same
   d. Drop to zero

4. (True or False) Distortion produced by overmodulation is filtered out in the diode detector. _____

5. Reversing the polarity of the detector diode causes the recovered signal to _____.
   a. Be inverted
   b. Be lowered in amplitude
   c. Be increased in amplitude
   d. Remain the same

ACTIVITY 3-5
LAB EXPERIMENT:
DOUBLE-SIDEBAND OPERATION

PURPOSE

To demonstrate the operation of a balanced modulator in producing a double-sideband signal.

MATERIALS

In addition to the 2206 function generator circuit which you built previously and all of the test instruments used earlier, you will need the following components:

Qty.
1 741 IC op amp
1 1496 balanced modulator IC
2 100-Ω resistors
3 560-Ω resistors
3 1-kΩ resistors
2 3.3-kΩ resistors

(continued)

2 4.7-kΩ resistors
1 6.8-kΩ resistor
2 0.1-μF capacitors
1 100-μF electrolytic capacitor
1 10-kΩ
1 100-kΩ

INTRODUCTION

Double-sideband (DSB) modulation refers to amplitude modulation but with the carrier removed. The modulator suppresses the carrier but produces full upper and lower sidebands as may be created by the modulating signal. Single-sideband (SSB) operation refers to amplitude modulation with the carrier suppressed but with one sideband removed. All of the intelligence is carried in either the upper or the lower sideband. SSB operation has the advantage of requiring one-half the spectrum space required by a normal AM or DSB signal.

The key circuit in generating a DSB or SSB signal is the balanced modulator. This circuit produces AM but suppresses the carrier, leaving only the two sidebands in the output. For SSB operation, a highly selective filter is normally used to eliminate one of the sidebands.

A variety of different balanced modulator circuits are used to suppress the carrier. One of the most effective and widely used is the differential amplifier balanced modulator. One such popular integrated circuit is the 1496. It can produce carrier suppression up to 65 dB and has a useful gain up to 100 MHz. It uses specially connected differential amplifiers to cancel the carrier.

In this experiment you will demonstrate the use of the 1496 to produce a DSB signal. You will also show how this IC can be used to produce conventional amplitude modulation.

PROCEDURE

1. Construct the circuit shown in Fig. 3-13. The 2206 integrated circuit function generator will serve as the carrier source. You will use a 741 op amp inverter to buffer the output signal from the function generator. The output of the 741 op amp becomes the carrier input to the 1496. The modulating signal is supplied by an external function generator. Be particularly careful in constructing this circuit, because there are many components and wiring mistakes are easy to make. Once you have built the circuit, double-check your wiring before applying power.

2. While monitoring the carrier signal at the op amp output on pin 6, adjust the 10-kΩ pot for a carrier amplitude of 2 V_{p-p}. By using the 100-kΩ pot on the 2206, set the carrier frequency to 20 kHz, and then adjust the modulating signal input from the external function generator for a frequency of 400 Hz with an amplitude of 0.5 V.

3. Connect the oscilloscope to pin 6 of the 1496 to observe the double-sideband output signal. The classical DSB signal pattern should appear. Adjust the 100-kΩ balance pot on the 1496 so that the arm is approximately in the center of its rotational range. Then fine-tune the pot adjustment until the positive and negative peaks of the DSB waveform are equal.

4. Readjust the 100-kΩ pot on the 2206 function generator to set the carrier frequency to approximately 6 kHz. The adjustment should not change the shape of the DSB output signal, but you should be able to see the carrier cycles more clearly. Next, by using the scope hori-

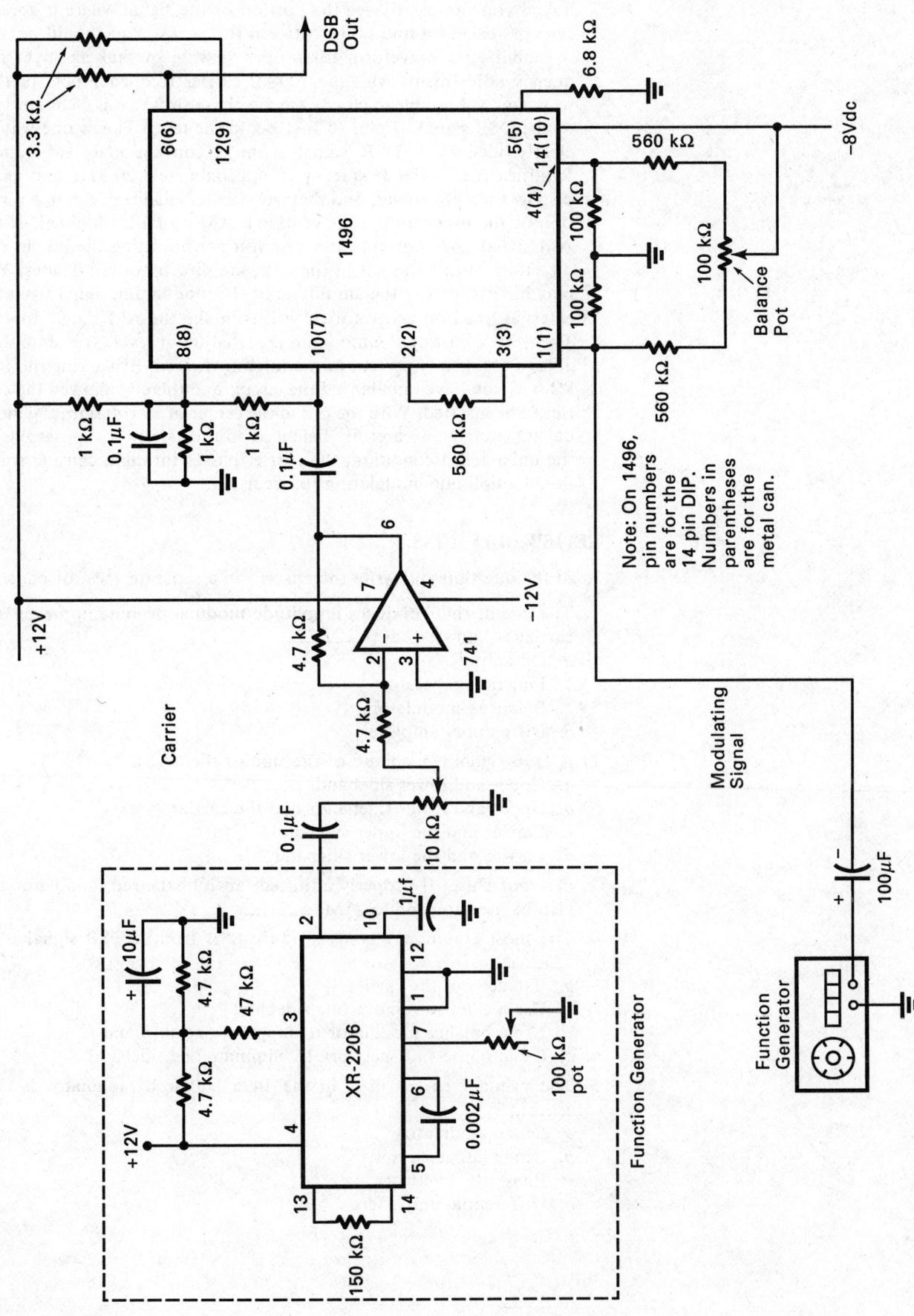

Fig. 3-13 Circuit for demonstrating SSB.

27

zontal and vertical gain and triggering controls, expand the DSB signal so you can clearly see the portion of the signal where it goes to zero or reaches a null point between the peaks. You should be able to stabilize the waveform for proper viewing by making slight frequency adjustments with the 100-kΩ carrier frequency pot. In that way, you will be able to clearly see the phase shift that is characteristic of the DSB signal. (Refer to Fig. 2-9 in the text.) This is one way to clearly identify a DSB signal from a conventional AM signal.

5. Readjust the carrier frequency to approximately 20 kHz and again display the DSB signal, and then rotate the balance pot in one direction or the other until a conventional AM signal is obtained. If the AM signal does not appear when you are adjusting the pot in one direction, adjust the pot in the opposite direction until it does. You may have to reduce the amplitude of the modulating signal from the external function generator as well to make the AM signal appear. Then, by adjusting the modulating signal input level, you should be able to obtain any percent of modulating desired. By varying the 100-kΩ pot, you have unbalanced the circuit and thereby allowed the carrier to be inserted. With the pot in the center of its rotation, the internal IC circuitry is carefully balanced so the carrier is suppressed. In the unbalanced condition, the carrier passes through, causing traditional amplitude modulation to occur.

REVIEW QUESTIONS

Read the question and write the answer on a separate sheet of paper.

1. The circuit that produces amplitude modulation but suppresses the carrier is known as a(n) _____.
   a. Op amp
   b. Function generator
   c. Balanced modulator
   d. Differential amplifier

2. A DSB signal is made up of the sum of the _____.
   a. Upper and lower sidebands
   b. Upper and lower sidebands and the carrier
   c. Carrier and the upper sideband
   d. Carrier and the lower sideband

3. (True of False) If properly adjusted, an IC balanced modulator can also be used to produce AM. _____

4. The most common way to produce SSB from a DSB signal is to _____.
   a. Balance out the carrier
   b. Use a filter to remove one sideband
   c. Use a balance modulator to suppress one sideband
   d. Use a phase shift network to eliminate one sideband

5. The primary circuit used in the 1496 balanced modulator is a(n) _____.
   a. Diode modulator
   b. Diode lattice circuit
   c. Phase-shift network
   d. Differential amplifier

CHAPTER 4

Angle Modulation

ACTIVITY 4-1
TEST: ANGLE MODULATION

Read the question and write the answer on a separate sheet of paper.

1. When PM is used, the frequency deviation is proportional to what characteristics of the modulating signal? _____
   a. Frequency
   b. Amplitude
   c. Both of the above
   d. Neither of the above

2. In order for a phase modulator to produce FM, which of the following must occur? _____
   a. Modulating signal must vary.
   b. Modulating signal must be constant.
   c. Carrier frequency must vary.
   d. Carrier amplitude must vary.

3. Which of the following is *not* a benefit of FM over AM? _____
   a. Excellent noise immunity
   b. Minimum use of spectrum space
   c. Interfering signal rejection due to capture effect
   d. Higher transmitter efficiency

4. How many significant sideband pairs are generated by an FM broadcast station with a maximum deviation of 75 kHz and a maximum modulating frequency of 15 kHz? _____
   a. 4   c. 8
   b. 6   d. 12

5. The FM sound transmitter of a TV broadcast station has a maximum derivation of 25 kHz and a maximum modulating frequency of 12.5 kHz. What is the bandwidth of the resulting signal? _____
   a. 25 kHz
   b. 50 kHz
   c. 75 kHz
   d. 100 kHz

Note: Use the formula for bandwidth given in the text.

6. If the maximum allowed deviation is 5 kHz and the maximum modulating signal is 3 kHz, the deviation ratio is _____.
   a. 0.6
   b. 0.8
   c. 1.5
   d. 1.67

7. In FM systems, high-frequency noise is minimized by using _____.
   a. High-frequency filtering
   b. Preemphasis
   c. Noise clippers
   d. Limiter amplifiers

8. What circuit must be used between the modulating signal and the phase modulator to produce FM? _____
   a. Low-pass filter
   b. High-pass filter
   c. Preemphasis circuit
   d. Amplitude limiter

9. Indirect FM is also known as _____.
   a. Angle modulation
   b. Amplitude modulation
   c. Phase modulation
   d. Rate-of-change modulation

10. What is the bandwidth of an FM signal produced by a modulator with a maximum deviation of 5 kHz and a maximum modulating signal of 3 kHz? Note: Use the bandwidth formula given in Activity 4-3. _____
    a. 8 kHz
    b. 12 kHz
    c. 16 kHz
    d. 24 kHz

11. During FM, the carrier amplitude in the frequency domain _____.
    a. Varies with modulation
    b. Remains constant
    c. Is suppressed
    d. Is phase-reversed

12. A TV sound transmitter has a maximum deviation of 25 kHz, and the actual deviation is 18 kHz. The percent modulation is _____.
    a. 13.9
    b. 45
    c. 68
    d. 72

13. Which of the following is *not* a typical FM application? _____
    a. Mobile radio
    b. Citizens band radio
    c. VCR
    d. Cellular telephone

14. The spacing between FM broadcast stations is _____ kHz.
    a. 150
    b. 175
    c. 200
    d. 250

15. In NBFM, *m* is made less than _____.

    a. $\pi/2$

    b. 1.667

    c. 2

    d. π

ACTIVITY 4-2
LAB EXPERIMENT:
FREQUENCY MODULATION

PURPOSE

To demonstrate frequency modulation with a voltage-controlled oscillator.

MATERIALS

You will need the 2206 function generator circuit you previously constructed plus an oscilloscope, a frequency counter, and an external function generator. You will also need a 10-μF capacitor and 100-kΩ resistor.

INTRODUCTION

In frequency modulation, the modulating signal varies the frequency of the carrier. The carrier amplitude remains constant, but the increasing and decreasing amplitude of the modulating signal causes the carrier to deviate from its center frequency. The amount of deviation is a function of the amplitude of the modulating signal. The rate of deviation is proportional to the frequency of the modulating signal.

There are many different ways to produce frequency modulation. Today, one of the easiest ways is simply to apply a modulating signal to a voltage-controlled oscillator (VCO), and numerous IC VCOs are available. The 2206 function generator IC you have used in past experiments uses a VCO as the primary signal-generating circuit. This oscillator can easily be frequency-modulated simply by applying the appropriate modulating signal. In this experiment you will demonstrate frequency modulation with the 2206.

PROCEDURE

1. The circuit for this experiment is shown in Fig. 4-1 on the next page. This is the same 2206 function generator you have used previously, but note that you will add a 10-μF capacitor and 100-kΩ resistor to pin 7. To the capacitor you will connect an external function generator to serve as the modulating signal.

2. Apply power to the circuit. Reduce the modulating signal amplitude from the external function generator to zero, and then observe the carrier output at pin 2 of the 2206. Adjust the 100-kΩ pot on the 2206 for a frequency of approximately 30 kHz. Set the horizontal sweep controls on the oscilloscope to display approximately three cycles of the carrier wave.

3. Slowly increase the amplitude of the modulating signal from the external function generator, and set the frequency of the signal to approximately 200 Hz. As you increase the modulating signal ampli-

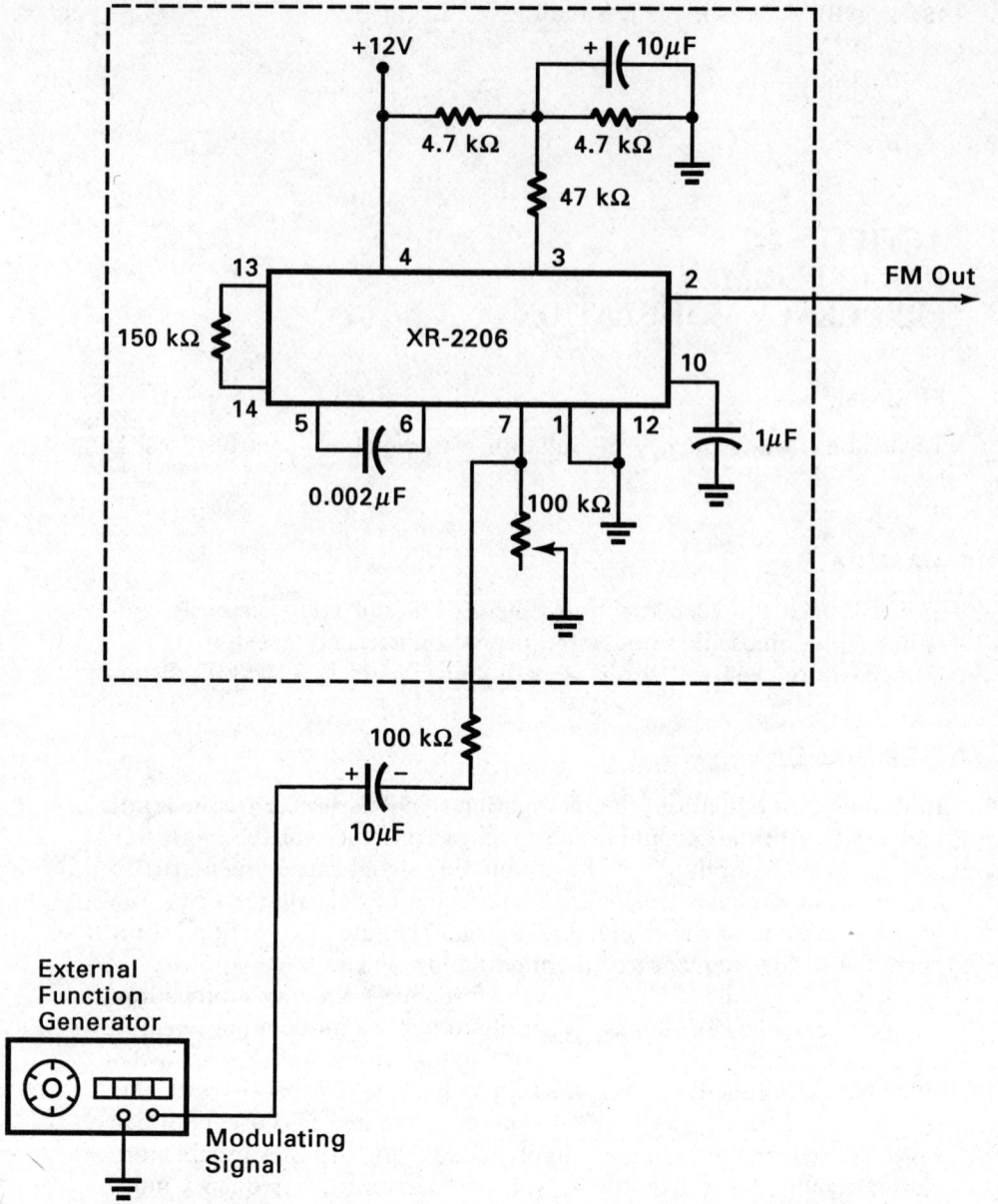

Fig. 4-1 Frequency modulation of the 2206.

tude, note the display on the oscilloscope. You should see the carrier wave begin to "vibrate," which indicates a change in frequency with the modulating signal. You should observe a waveform that looks approximately like the one shown in Fig. 4-2. Continue to slowly increase the modulating signal amplitude and note the effect on the carrier wave. If you have difficulty obtaining a stable display, use the triggered sweep function on the oscilloscope. Manipulate the trigger and horizontal frequency controls until a stable waveform is obtained. What you are seeing is the carrier frequency being instantaneously changed by the modulating signal. While varying the amplitude of the modulating signal and observing the resulting FM output, determine the relationship between the frequency deviation and the modulating signal amplitude.

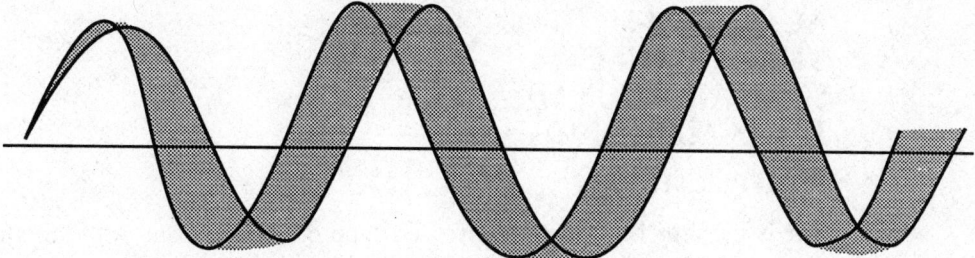

Fig. 4-2 How FM looks on an oscilloscope.

4. Set the amplitude of the modulating signal at the 10-μF capacitor to 2 $V_{\text{p-p}}$, and then observe the carrier output on the oscilloscope. You will see the waveform shown in Fig. 4-2, indicating frequency deviation. Now, while observing the waveform, vary the frequency of the modulating signal. Note the effect on the waveform. How does the deviation change with respect to the modulating signal frequency?

REVIEW QUESTIONS

Read the question and write the answer on a separate sheet of paper.

1. Which of the following is the most correct statement with regard to the relationship between the carrier frequency deviation and the amplitude of the modulating signal? _____
   a. Decreasing the modulating signal amplitude increases the carrier deviation.
   b. Decreasing the modulating signal amplitude decreases carrier deviation.
   c. Carrier deviation is not affected by the amplitude of the modulating signal.
   d. The frequency deviation varies in direct proportion to the modulating signal frequency.

2. Varying the frequency of the modulating signal causes the frequency deviation of the carrier to _____.
   a. Increase
   b. Decrease
   c. Remain the same
   d. Drop to zero

3. The type of modulation produced by the VCO in the 2206 IC is _____.
   a. Frequency modulation
   b. Phase modulation
   c. Indirect FM
   d. FSK

4. What is the deviation ratio of an FM system in which the maximum permitted frequency deviation is 10 kHz and the maximum modulating frequency is 3 kHz? _____
   a. 0.3
   b. 1
   c. 3
   d. 3.33

5. The number of sidebands produced by a sine wave carrier being modulated by a single-frequency sine wave tone is _____.
   a. 1   c. 4
   b. 2   d. Infinite

ACTIVITY 4-3
STUDY PROJECT:
SOME ADDITIONAL
FACTS ABOUT FM

FM APPLICATIONS

FM is perhaps the most widely used type of modulation. Some of the more common applications are listed below.

1. FM radio broadcasting
2. TV sound broadcasting
3. Cellular telephones
4. Cordless telephones
5. Marine VHF radio
6. Satellite broadcasting and relay
7. Cable TV
8. Mobile radio (police, fire, taxi, etc.)
9. Amateur radio (VHF, UHF)
10. VCRs
11. Some types of computer modems
12. Some types of radar
13. Some military radios
14. Wireless microphones

SPECTRUM OF THE FM BROADCAST BAND

The FM radio broadcast band extends from 88 to 108 MHz, and stations are spaced every 200 kHz from 88.1 to 107.9 MHz. Each station is allocated a 150-kHz space in the spectrum with 25-kHz guard bands on each side. See Fig. 4-3.

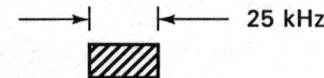

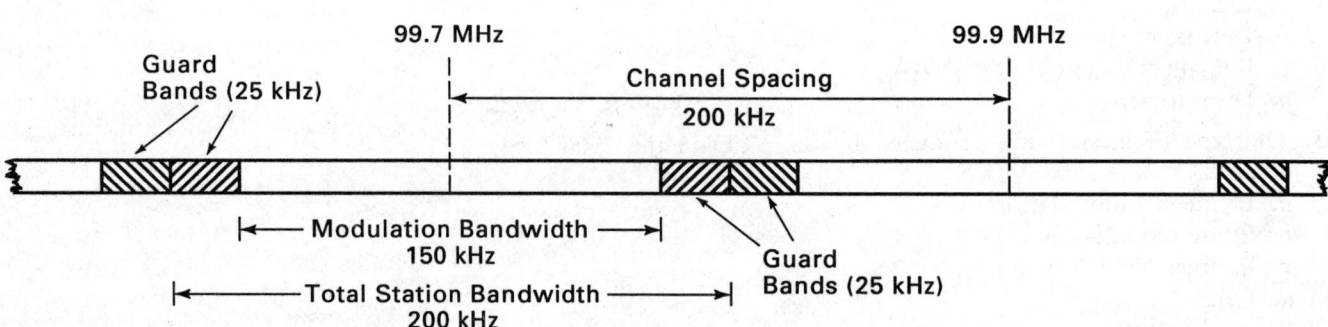

Fig. 4-3 Spectrum of broadcast FM radio.

NARROWBAND FM

To minimize the spectrum space occupied by an FM signal, the modulation index can be restricted, which will reduce the number of significant sidebands that occur. In narrowband FM (NBFM), the modulation index is held to less than $\pi/2$. In this way, only a few pairs of significant sidebands occur, thereby reducing signal bandwidth. For example, a

34

maximum deviation of 3 kHz with a maximum modulating frequency of 3 kHz produces a modulation index of 3 kHz/3 kHz = 1. According to the Bessel function table, only three significant pairs of sidebands are produced. The bandwidth can be estimated from the formula:

$$BW = 2f_m \times \text{number of significant sidebands}$$

where f_m is the maximum modulating frequency.

In the above example,

$$BW = 2(3 \text{ kHz})(3) = 18 \text{ kHz}$$

Another way to compute FM bandwidth is with the formula:

$$BW = 2(f_d + f_m)$$

where f_d is the maximum deviation and f_m is the maximum modulating frequency. This formula takes into account only the sidebands that contribute 96 percent of the total signal power.

Using the preceding example, the bandwidth is:

$$BW = 2(3 \text{ kHz} + 3 \text{ kHz}) = 12 \text{ kHz}$$

This bandwidth formula is a more accurate indicator in practical applications than the formula given earlier.

CHAPTER | 5

Frequency Modulator Circuits

Read the question and write the answer on a separate sheet of paper.

1. To use a PN junction diode as a voltage variable capacitor, the diode must be _____.
   a. Forward-biased
   b. Reverse-biased

2. Increasing the bias voltage on a varactor causes the varactor capacitance to _____.
   a. Decrease
   b. Increase

3. A VVC in parallel with a parallel LC-tuned circuit in an oscillator has its bias voltage decreased by the modulating signal. The oscillator frequency _____.
   a. Increases
   b. Decreases

4. (True or False) Crystal oscillators cannot be frequency-modulated. _____

5. An oscillator that can be frequency-modulated is often referred to as _____.

6. PM is referred to as _____ FM.

7. Shifting the phase of a carrier in accordance with the modulating signal produces _____.

8. (True or False) A VVC can be used to produce PM. _____

9. A phase modulator produces a deviation of 800 Hz on a carrier of 6.3 MHz. Frequency multipliers of three and five are used. The final output frequency is _____ MHz with a deviation of _____.

10. Foster-Seeley discriminators and ratio detectors convert frequency changes into _____ variations which reproduce the original intelligence signal.

11. In a pulse-averaging discriminator output pulses from a one-shot are averaged in a(n) _____.

36

12. A quadrature demodulator converts frequency changes into pulse _____ changes that translate into amplitude variations.

13. The output of the phase detector in a quadrature demodulator is passed through a(n) _____ to recover the original intelligence signal.

14. The best FM demodulator is generally considered to be the _____.

15. Name the FM demodulators that require limiters. _____

16. The output of a PLL FM demodulator is taken from the _____.
   a. Phase detector
   b. VCO
   c. Low-pass filter

ACTIVITY 5-2
LAB EXPERIMENT:
PHASE-LOCKED LOOP OPERATION

PURPOSE

To demonstrate the operation and determine the lock and capture ranges of a phase-locked loop.

MATERIALS

Qty.
1   Oscilloscope (dual trace)
1   Frequency counter
1   Function generator (sine wave)
1   Power supply (± 12 V dc)
1   565 IC PLL
1   0.001-μF capacitor
1   0.01-μF capacitor
1   0.1-μF capacitor
1   10-μF electrolytic
2   1-kΩ resistors
1   2.7-kΩ resistor

INTRODUCTION

A phase-locked loop (PLL) is an electronic feedback control system used in a variety of applications. It consists of an error detector, a loop filter, and a voltage-controlled oscillator (VCO). See Fig. 5-1 on the next page.

A phase-locked loop will track or follow the frequency of an input signal. The input signal frequency is compared to the VCO frequency in the phase detector, and an error signal is generated. This error signal from the phase detector is filtered into a dc voltage which controls the VCO. Changes in the input frequency develop an error signal that will cause the VCO's output frequency to change in such a direction as to exactly follow the input signal frequency. Frequency changes of the input are interpreted as phase changes which cause the VCO to readjust. The VCO output frequency is equal to the PLL input frequency, but there is a phase shift between the input and VCO signals. It is this phase shift that generates the error signal and the dc voltage that keeps the phase-locked loop locked. In this experiment, you will demonstrate PLL operation by using the popular 565 IC phase-locked loop.

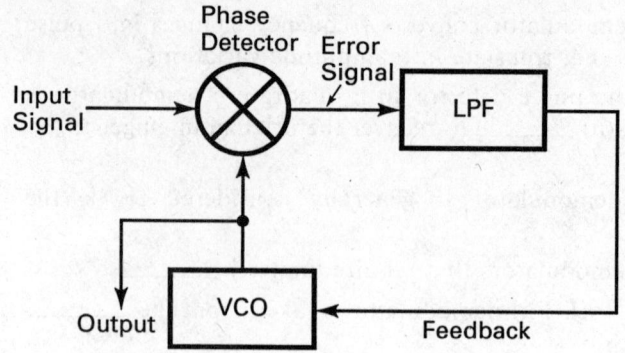

Fig. 5-1 Phase-locked loop.

In the 565 integrated circuit phase-lock loop, the VCO is of the multivibrator type and produces both triangular and square wave outputs. The frequency is determined by external values of resistance and capacitance. The approximate free-running frequency is computed with the formula:

$$f_o = \frac{1.2}{4R_1C_1}$$

Refer to Fig. 5-2, which shows the 565 and the related components.

The frequency range over which the VCO is capable of tracking the input signal is known as the lock range. It is primarily a function of the range of the VCO, and it is centered around the free-running frequency of the VCO. For the 565 PLL, it is computed with the formula:

$$f_L = \frac{\pm 8 f_o}{V_s}$$

where f_L is one-half the lock range, f_o is the free-running VCO frequency, and V_s is the total supply voltage. In this circuit, there is a $+V_{cc}$ supply and a $-V_{EE}$ supply for a total of $12 + 12 = 24$ V. The total lock range is $2f_L$.

The capture range is the range of frequencies over which the PLL will lock onto an input signal. When an input signal is applied to the VCO, it must be within the capture range in order for the circuitry to "capture" the signal. A signal whose frequency is outside the capture range will simply cause the PLL to remain unlocked, and the VCO will operate at its free-running frequency. The capture range is usually narrower than the lock range, which causes the PLL to act like a very selective band-pass filter.

In this experiment, you will demonstrate PLL operation and measure the phase shift between the input signal and the VCO during the locked state. You will also demonstrate how the VCO output tracks a variable-frequency input. Finally, you will calculate and measure the free-running frequency and the lock range of the 565 PLL.

PROCEDURE

1. Connect the circuit shown in Fig. 5-2. An external function generator with variable-frequency capability is connected as the PLL input. Reduce the amplitude to zero for now.

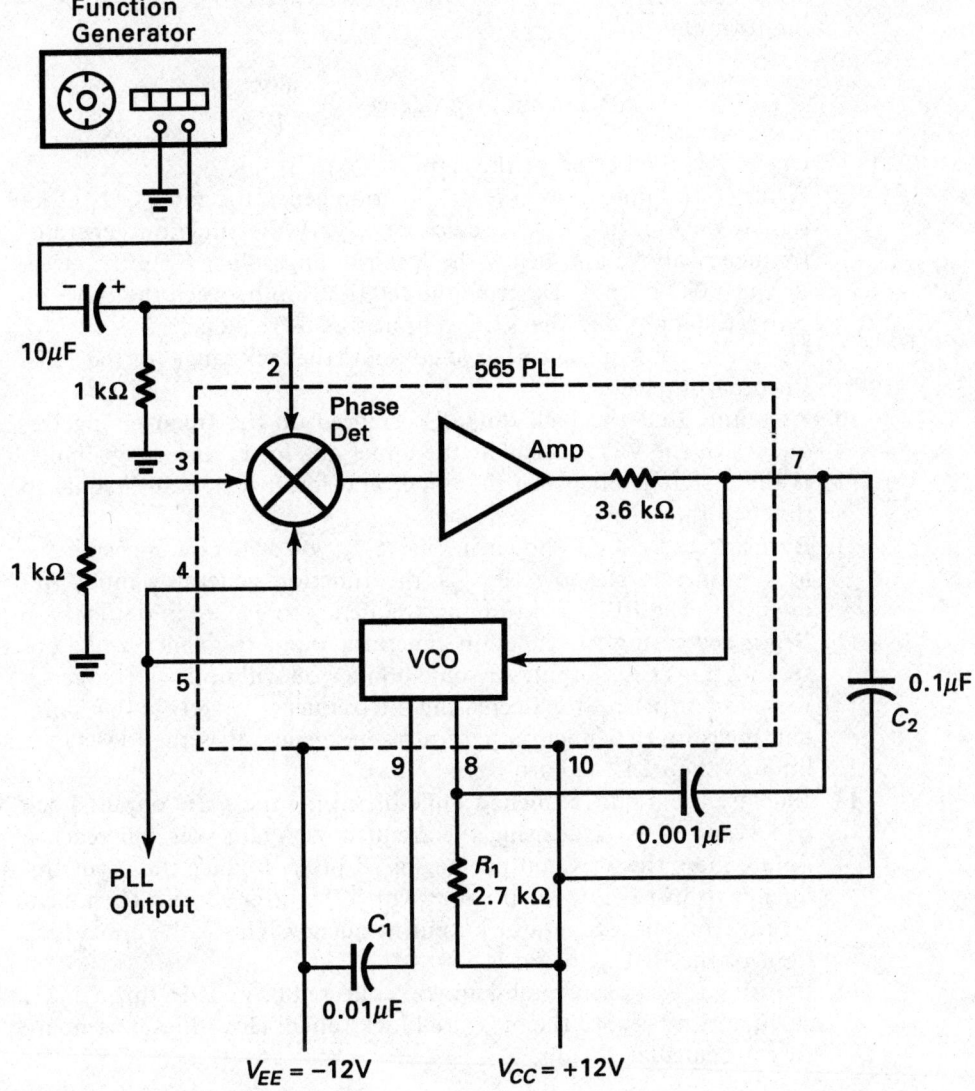

Fig. 5-2 Phase-locked loop circuit.

2. Use the formula for the free-running frequency of the VCO, given earlier, to calculate that frequency by using the values of R_1 and C_1 in Fig. 5-2.

3. Apply power to the circuit shown in Fig. 5-2. By using a frequency counter, measure the output frequency of the VCO at pin 5 on the 565 IC. Observe the output signal with an oscilloscope.

4. Compare your computed and measured values of free-running frequency and explain any difference.

5. Adjust the function generator input to the PLL input for an amplitude of 1 V each peak to peak of sine wave. Set the external function generator for a frequency approximately equal to the free-running frequency of the VCO.

6. Use an oscilloscope and/or frequency counter to measure the function generator output frequency and the VCO output frequency at pin 5. Are the two frequencies the same? _____ Why or why not?

7. Use a dual-trace oscilloscope to display the function generator and VCO output signals simultaneously. Measure the amount of phase shift between the two. You can do that by measuring the time shift

t between corresponding parts of the two waveforms and then using the formula:

$$\text{Phase shift, in degrees} = \frac{360t}{T}$$

where T is the period of the signals.

8. While continuing to observe the function generator and VCO output signals on the dual-trace oscilloscope, vary the function generator frequency above and below the free-running value. Note the effect on the VCO output. Describe the relationship between the function generator input and the VCO output signal frequencies.

9. Use the formula given earlier to compute the lock range for the PLL. Record your value.

10. Assuming that the lock range is centered on the free-running frequency of the VCO, calculate the upper and lower lock range limits. Is the difference between the upper and lower lock values equal to the lock range?

11. By using the circuit shown in Fig. 5-2, you will now measure the lock range. First, however, set the function generator input frequency to the PLL free-running frequency to ensure initial lock.

12. Begin decreasing the function generator input frequency while observing the VCO output. At some point, you will notice a frequency variation or jitter. Stop decreasing the frequency exactly at that point and measure the function generator frequency. It is the lower lock limit of the PLL. Record it.

13. Increase the input frequency while observing the VCO ouput. Lock will occur. Keep increasing the frequency. Again you will reach a point where the VCO output begins to jitter. Reduce the input frequency to just below the point at which the jitter ceases. Then measure the function generator output frequency. This is the upper lock limit of the PLL. Record it.

14. By using the experimental data you collected in steps 16 through 21, calculate and record the measured lock range. How does it compare to your calculated value?

REVIEW QUESTIONS

Read the question and write the answer on a separate sheet of paper.

1. Varying the PLL input frequency causes the PLL output to _____.
   a. Track the input
   b. Remain constant at the free-running frequency
   c. Vary inversely
   d. Drop to zero

2. The error signal is generated by what characteristic of the input and VCO signals? _____
   a. Frequency
   b. Phase shift
   c. Amplitude difference
   d. Rise time

3. What is the relationship between the capture f_c and lock f_L ranges of the PLL? _____
   a. $f_c = f_L$
   b. $f_c > f_L$
   c. $f_c < f_L$
   d. $f_c = 2f_L$

4. To increase the free-running or center frequency of the PLL, what changes should be made in R_1 and/or C_1? ___ e ___

a. Increase R_1

b. Increase C_1

c. Decrease R_1

d. Decrease C_1

e. Both a and b

f. Both c and d

5. If the input frequency to the PLL is outside the capture and lock ranges, the VCO output is the ___ b ___.

a. Upper lock frequency

b. Free-running frequency

c. Lower lock frequency

d. Input frequency

ACTIVITY 5-3
LAB EXPERIMENT:
FREQUENCY DEMODULATION
WITH A PHASE-LOCKED LOOP

PURPOSE

To demonstrate the operation of a phase-locked loop as a demodulator for FM signals.

MATERIALS

In addition to an oscilloscope, frequency counter, function generator, and power supply, you will need the 2206 FM modulator you built in Activity 4-1 and the 565 PLL circuit you used in Activity 5-1. You will also need a 0.001-μF capacitor and a 10-kΩ pot.

INTRODUCTION

Perhaps the best demodulator circuit for FM is the phase-locked loop. Because of its excellent fidelity, wide range response, and noise suppression, the PLL produces a clean and faithful reproduction of the original modulating signal. Because PLLs are available in low-cost IC form, they are widely used for this purpose, particularly in critical applications.

To use the PLL as a frequency demodulator, you apply the FM signal to the phase detector input. A recovered signal is taken from the output of the loop low-pass filter (LPF). See Fig. 5-3.

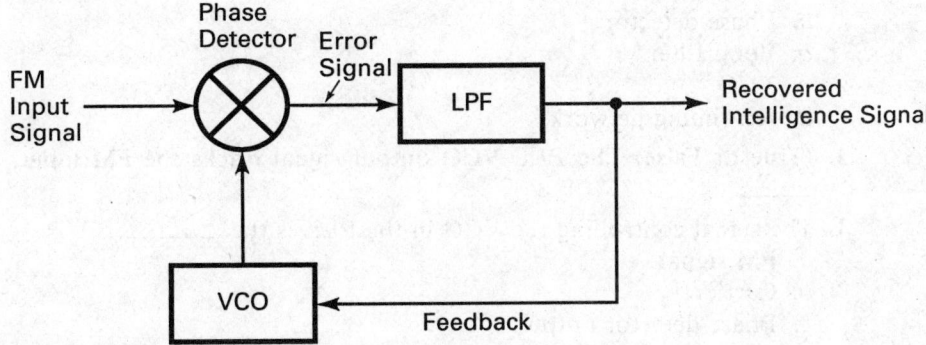

Fig. 5-3 Phase-locked loop used as an FM demodulator.

As the carrier varies in accordance with the modulating signal, it produces a varying phase when compared to the VCO. The error signal developed then forces the VCO output to follow the FM input. Remember that, in order for the PLL to remain locked, the two inputs to the phase detector must be the same The VCO output will track the FM input.

In order for the VCO to exactly duplicate the input FM signal, the signal applied to the VCO control input from the loop filter must be the same as the modulating signal. The output of the loop filter is, therefore, the recovered intelligence signal.

In this experiment you will use the 565 IC PLL to demonstrate FM demodulation.

PROCEDURE

1. The complete circuit for this experiment is shown in Fig. 5-4. It is made up of the 2206 IC frequency modulator you used in the preceding experiment, and the demodulator circuit itself is the 565 PLL you used in that experiment. Connect the output of the frequency modulator to the 565. In the PLL circuit, replace the 0.01-μF capacitor with a 0.001-μF capacitor. Replace the 2.7-kΩ resistor R_1 with a 10-kΩ pot.
2. Connect a frequency counter and oscilloscope to the 2206 output at pin 2. Reduce the modulating input signal to zero. Adjust the 100-kΩ pot for a carrier frequency of 30 kHz.
3. Connect the frequency counter and oscilloscope to the PLL output at pin 5 on the 565. Adjust the 10-kΩ pot for a frequency of 30kHz. At this time, the 565 PPL should be locked to the 2206 output; check that by observing the 565 output at pin 5 while varying the 100-kΩ pot on the 2206. The output should track. Readjust the 100-kΩ pot for an output of 30 kHz.
4. Apply a modulating signal from the external function generator to the 2206 FM generator. Use a frequency of 200 to 400 Hz. While observing the output of the 2206 at pin 2, increase the amplitude of the modulating signal until you obtain an FM signal.
5. Observe the PLL loop filter output at pin 7 to see the recovered signal. Compare that signal to the modulating signal applied to the 2206.
6. While observing the recovered output at pin 7 on the 565 PLL, vary both the frequency and amplitude of the modulating signal. Note the effect on the recovered output.

REVIEW QUESTIONS

Read the question and write the answer on a separate sheet of paper.

1. The recovered output from a PLL demodulator appears at the output of the ___b___.
   a. Phase detector
   b. Loop filter
   c. VCO
   d. RC timing network
2. (True or False) The PLL VCO output signal tracks the FM input. ___T___

3. The signal controlling the VCO in the PLL is the ___d___.
   a. FM signal
   b. Carrier
   c. Phase detector output
   d. Original modulating signal

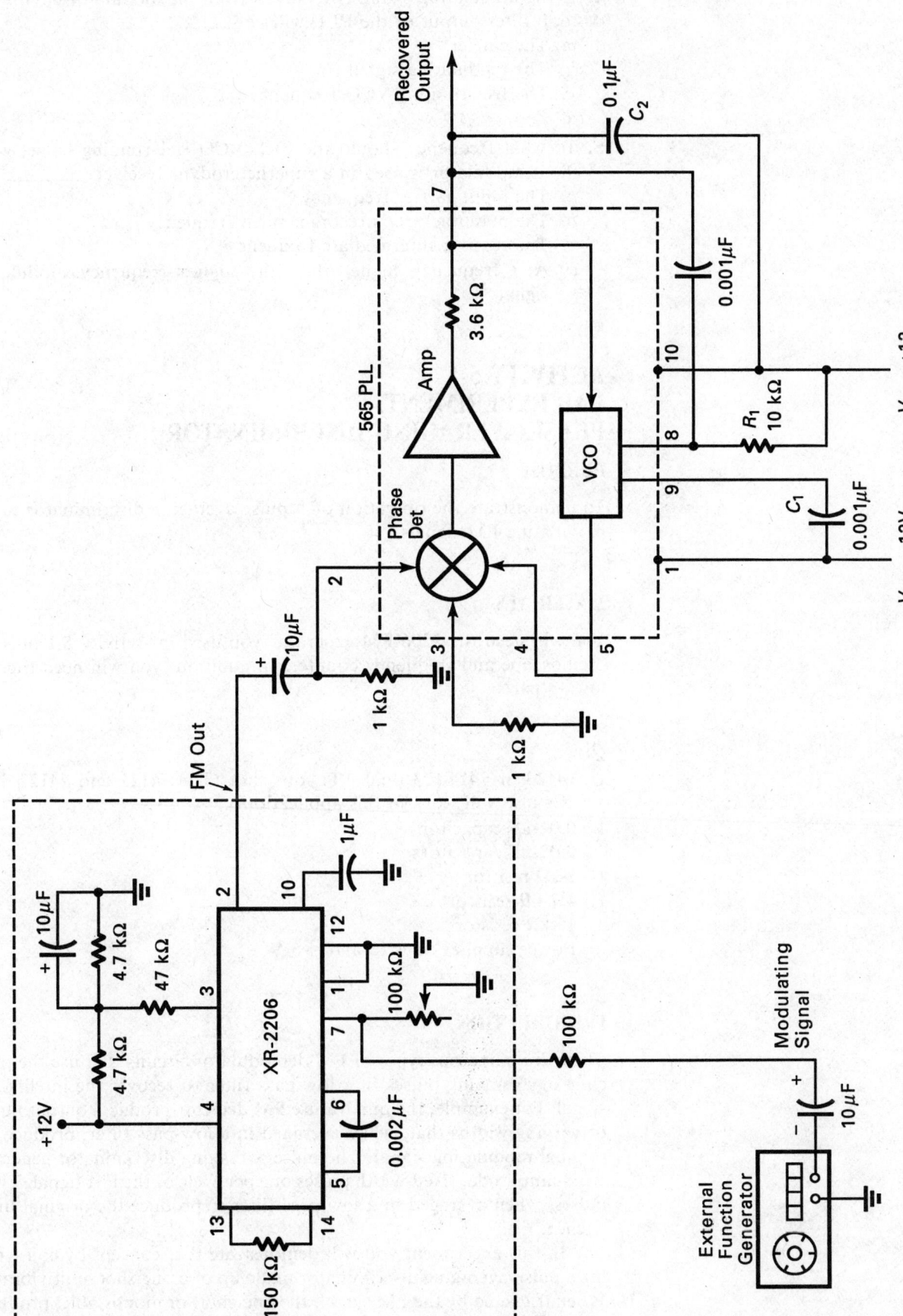

Fig. 5-4 Using a PLL for FM demodulation.

4. If the modulating signal is removed from the modulator circuit, the loop filter output of the PLL will be _____d_____.
   a. The carrier
   b. The modulating signal
   c. The free-running VCO frequency
   d. Zero

5. To what frequency should the PLL VCO free-running be set when the demodulator is used in a superheterodyne receiver? _____c_____
   a. The input carrier frequency
   b. The manufacturer's recommended frequency
   c. The receiver intermediate frequency
   d. Any frequency higher than the highest-frequency modulating signal

ACTIVITY 5-4
LAB EXPERIMENT:
PULSE-AVERAGING DISCRIMINATOR

PURPOSE

To demonstrate the operation of a pulse-averaging discriminator in demodulating FM signals.

MATERIALS

You will need the 2206 FM generator you used in Activity 5-1 plus the oscilloscope and frequency counter. In addition, you will need the following parts:

Qty.
1 74123 or 74LS123 dual TTL one-shot (The 74121 and 74122 TTL ICs also will work in this application.)
1 0.01-μF capacitor
2 0.02-μF capacitors
1 1-kΩ resistor
1 4.7-kΩ resistor
2 15-kΩ resistors
   Power supplies of $+12$ and $+5$ V

INTRODUCTION

There are numerous types of FM demodulator circuits that use the principle of averaging pulses in a low-pass filter to recover the intelligence signal. For example, the quadrature FM detector produces output pulses of varying widths that, when averaged in a low-pass filter, produce the original modulating signal. The pulse-averaging discriminator generates fixed-amplitude, fixed-width pulses one per cycle of the FM signal. These pulses, when averaged in a low-pass filter, reproduce the original intelligence.

In this experiment you will demonstrate that concept by constructing a pulse-averaging discriminator made up of a one-shot multivibrator. When triggered by the FM signal, the one-shot, or monostable, produces one fixed-width pulse per cycle. The pulses are fed to a low-pass filter to reproduce the original signal.

PROCEDURE

1. Construct the experimental circuit shown in Fig. 5-5. The output from the 2206 function generator is taken from pin 11; it is a 5-V square wave compatible with the TTL logic circuitry in the 74123 one-shot. The signal is also frequency-modulated and follows exactly the sine wave output which appears at pin 2 on the 2206.

2. Connect the frequency counter to the 2206 output at pin 11, and then adjust the 100 kΩ pot for a center frequency of 20 kHz. While you do that, be sure to reduce the modulating signal input from the external function generator to zero.

3. Connect the oscilloscope output to pin 13 of the 74123 one-shot. You should observe a chain of pulses, and the pulse width should be approximately 15 μs. By using the 100-kΩ pot on the 2206, vary the frequency over a narrow range above and below the 20-kHz point. Note that the frequency of the pulses changes but the pulse width remains constant.

4. Again set the 2206 function generator for a center frequency of 20 kHz with the 100-kΩ pot, and then apply the modulating signal to the 2206 from the external function generator. Set the modulating signal frequency to 200 Hz and monitor the output of the 2206 at pin 11. You should see a frequency-modulated square wave. This is the signal applied to the one-shot.

5. Connect the oscilloscope to the output of the low-pass filter. You should observe the recovered 200-Hz sine wave. Vary the amplitude of the modulating signal and note the effect on the output.

6. While observing the demodulating output, increase the modulating signal frequency from 200 Hz slowly up to approximately 2 kHz. Note the effect on the output. How does the demodulated signal vary with frequency? Explain the reason for the action you observe.

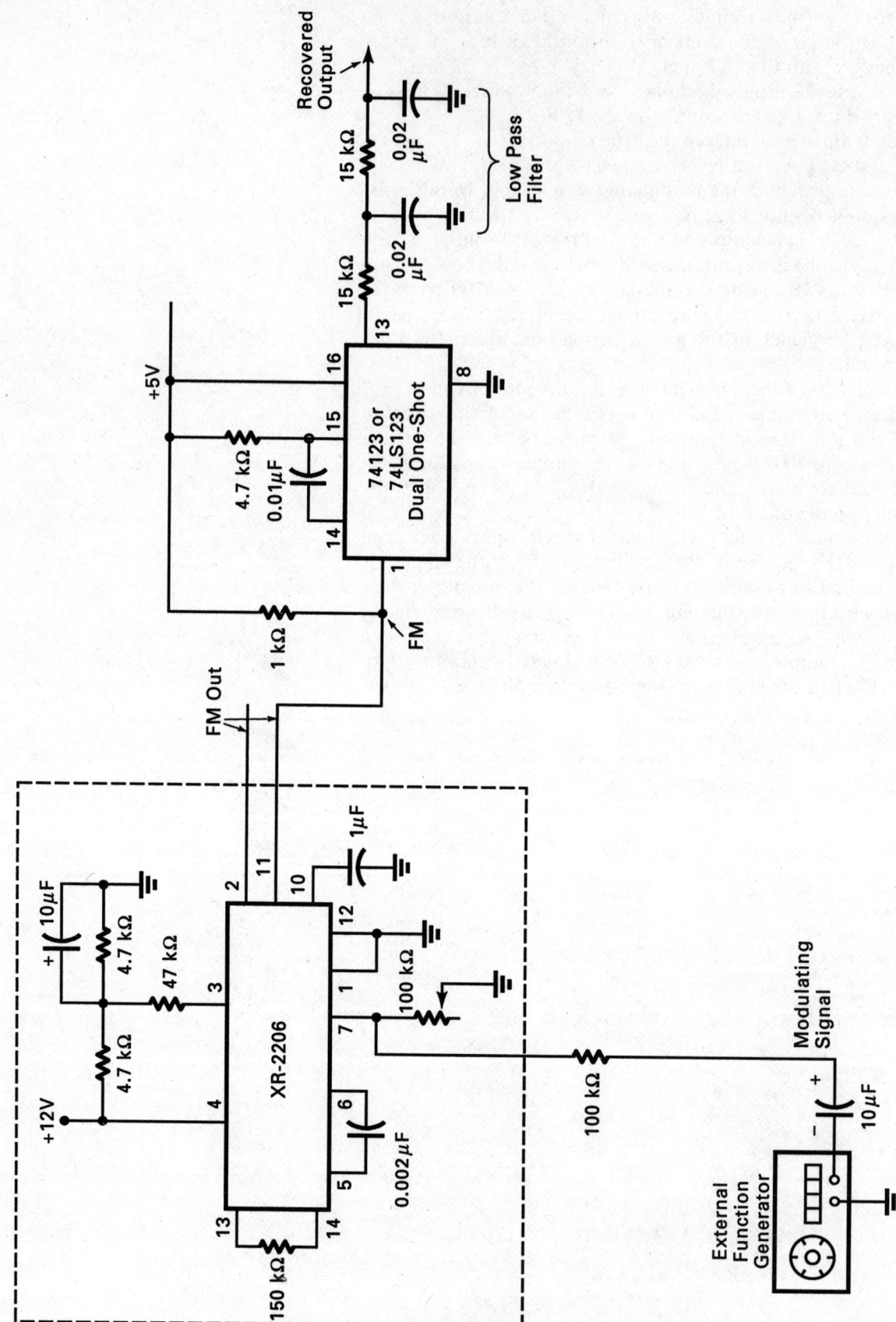

Fig. 5-5 Pulse-averaging discriminator or demodulator.

46

REVIEW QUESTIONS

Read the question and write the answer on a separate sheet of paper.

1. The circuit that generates fixed-width fixed-amplitude pulses is known as a(n) _d_____.
   a. Astable multivibrator
   b. Pulse generator
   c. Oscillator
   d. One-shot multivibrator

2. The circuit that smooths the pulses into the original intelligence signal is the _A____.
   a. Low-pass filter
   b. High-pass filter
   c. One-shot multivibrator
   d. Quadrature detector

3. Another FM demodulator that uses the principle of averaging pulses in a filter is the __C__.
   a. Phase-locked loop
   b. Foster-Seeley discriminator
   c. Quadrature detector
   d. Ratio detector

4. The amplitude of the recovered intelligence signal is directly proportional to __b__.
   a. Carrier frequency   c. Pulse amplitude
   b. Carrier deviation   d. Pulse width

5. As the frequency of the modulating signal increases, the recovered signal output amplitude __C__.
   a. Decreases
   b. Increases
   c. Remains the same
   d. Drops to zero

CHAPTER | 6

Radio Transmitters

ACTIVITY 6-1
TEST: RADIO TRANSMITTERS

Read the question and write the answer on a separate sheet of paper.

1. Which of the following is *not* a typical stage in a transmitter?
   _____
   a. Carrier oscillator
   b. Modulator
   c. AGC amplifier
   d. RF power amplifier

2. The amplifiers of class _____ are the most efficient.
   a. A
   b. AB
   c. B
   d. C

3. In a high-level AM transmitter, the final amplifier usually operates in which class? _____
   a. A
   b. AB
   c. B
   d. C

4. Which type of power amplifier do low-level AM and SSB transmitters use? _____
   a. Linear
   b. Push-pull
   c. Class C
   d. Feedback

5. The purpose of impedance-matching circuits and components is to _____.
   a. Minimize signal distortion
   b. Reduce spurious emissions
   c. Achieve maximum power output
   d. Set the frequency of operation

6. Which of the following is *not* a kind of impedance-matching device?

    _a._ L network   575-576
    _b._ Ferrite bead
    _c._ π network
    _d._ Balun

7. A toroid has a 15-turn primary and a 7-turn secondary. The load impedance is 72 Ω. The input or generator impedance is _____ Ω.
    _a._ 15.7
    _b._ 85.9
    _c._ 204.6
    _d._ 330.6

8. Speech-processing circuits are *not* used for _____.
    _a._ Improving fidelity
    _b._ Preventing overmodulation
    _c._ Restricting bandwidth
    _d._ Minimizing adjacent channel splatter

9. An FM transmitter has a 15-MHz carrier oscillator and frequency multipliers of 2, 3, and 5. The final output frequency is _____ MHz.
    _a._ 25
    _b._ 150
    _c._ 300
    _d._ 450

10. Compute the L and C values of an L network to match a 20-Ω final amplifier to a 300-Ω load at 50 MHz. _____
    _a._ $L = 28.3 \ \mu H, \ C = 33 \ pF$
    _b._ $L = 238 \ nH, \ C = 40 \ pF$
    _c._ $L = 169 \ nH, \ C = 20 \ pF$
    _d._ $L = 357 \ nH, \ C = 47 \ pF$

11. A bipolar transistor class C amplifier has a collector supply voltage of 24 V and a collector current of 0.16 mA. The power input is _____ W.
    _a._ 3.84
    _b._ 9.5
    _c._ 38.4
    _d._ 150

12. (True or False) A class C amplifier generates harmonics. __T__

13. A toroid auto transformer is used to match impedances between a stage with an 8-Ω output to a stage with a 175 input impedance. The turns ratio (N_p/N_s) is _____.
    _a._ 0.046
    _b._ 0.214
    _c._ 4.6
    _d._ 21.9

14. A transmitter has an output of 224 MHz, and the carrier oscillator operates at a frequency of 12.444 MHz. The frequency multiplier chain has a multiplication factor of _____.
    _a._ 8   _c._ 14
    _b._ 12   _d._ 18

15. Increasing the reverse bias on a class C amplifier causes the conduction angle or collector pulse current width to _____.
    _a._ Increase
    _b._ Decrease
    _c._ Remain the same

ACTIVITY 6-2
LAB EXPERIMENT:
CLASS C AMPLIFIERS
AND FREQUENCY MULTIPLIERS

PURPOSE

To demonstrate the operation and use as a frequency multiplier of a class C amplifier.

MATERIALS

Qty.

1   Function generator
1   Oscilloscope
1   Frequency counter
1   Digital multimeter
1   NPN transistor (2N3904, 2N4124, 2N4401, 2N2222, etc.)
1   2-mH inductor
1   0.02-μF capacitor
1   0.1-μF capacitor
1   10-kΩ resistor
1   Power supply for +5 V

INTRODUCTION

Many radio transmitters use class C amplifiers for power amplification, and that is particularly true of FM transmitters. Most AM transmitters or those transmitting SSB use linear amplifiers. Class C amplifiers are usually preferred because they are far more efficient than linears, but they distort the AM/SSB signals. They can, however, be used in FM transmitters, which means that they generate more output power vs. input power than any other kind of amplifier.

A transistor class C amplifier is one that conducts for less than one half-cycle (180°) of the input sine wave signal. Class C amplifiers conduct only during the half cycle when the emitter-base junction in the transistor is forward-biased. Most class C amplifiers use some form of base bias that keeps the transistor cut off until the input signal polarity and amplitude are high enough to cause it to conduct. Typically, the bias is set up so that conduction occurs during a 90° to 150° range. Since conduction is less than one half-cycle, the collector current is, of course, highly distorted. However, since a tuned circuit is connected in series with the collector, the output signal is a complete sine wave. The flywheel effect of the tuned circuit supplies the missing half-cycle.

Class C amplifiers are also used as frequency multipliers. By connecting a tuned circuit in the collector whose resonant frequency is some integral value of the input frequency, frequency multiplication occurs. That provides a simple and low-cost way to increase the signal frequency. In that way, stable low-frequency circuits can be used to generate the original signal, and multipliers are used to increase the signal to its final higher RF value.

The class C amplifier is a good frequency multiplier because of the distortion it generates. The distortion, of course, means that the output signal contains many harmonics. The tuned circuit in the collector is used to select the desired harmonic.

In this experiment, you will demonstrate a simple transistor class C amplifier and show its use as a multiplier.

PROCEDURE

1. Construct the class C amplifier circuit shown in Fig. 6-1. Initially connect the function generator, but be sure that the input amplitude is reduced to zero initially.

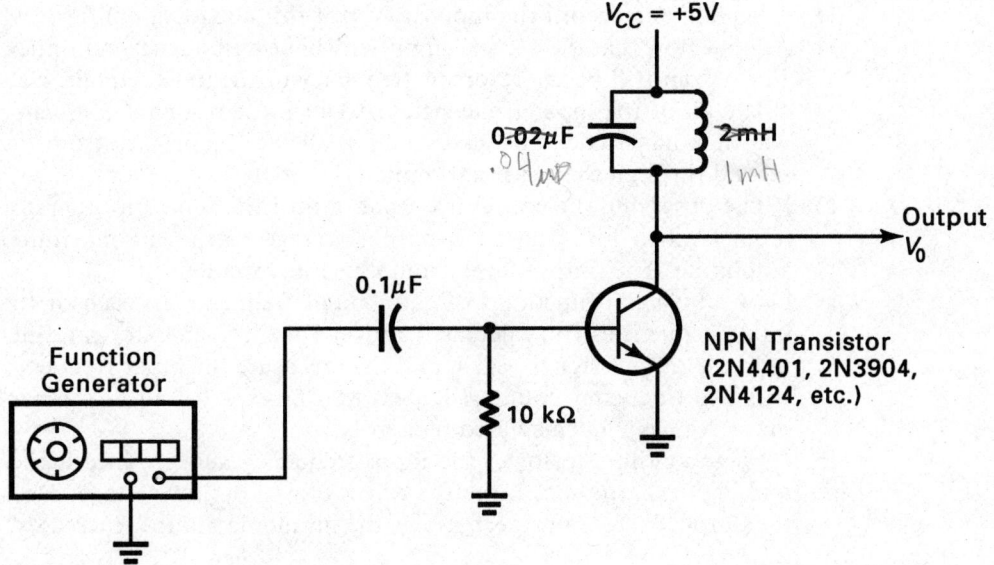

$V_{CC} = +5V$

0.02μF   .04μF   2mH   1mH

0.1μF

Output V_0

Function Generator

10 kΩ

NPN Transistor (2N4401, 2N3904, 2N4124, etc.)

Fig. 6-1 Class C amplifier.

2. Observe the tuned circuit in Fig. 6-1 and calculate the resonant frequency by using the values given. Record the frequency.

3. Connect an oscilloscope to the class C amplifier collector output. While observing the signal there, begin increasing the function generator input signal. Apply approximately 1.5 V_{p-p} to the input capacitor. The function generator frequency should be set to the resonant frequency value you computed in step 2. While observing the output, carefully tune the function generator frequency for maximum output voltage at the collector.

4. Measure and record the output frequency with the frequency counter. Your measured value should be close to the computed resonant frequency.

5. While observing the collector output on the oscilloscope, vary the function generator input voltage from zero until the output signal increases. At just the point where the output signal amplitude reaches its maximum output, stop increasing the input voltage.

6. By using a digital multimeter, measure the voltage across the 10-kΩ base resistor. Record the polarity and voltage.

7. While monitoring the dc voltage across the 10-kΩ resistor, continue increasing the input voltage from the function generator. Increase it to a value of about 4 V_{p-p} and note how the bias voltage changes. Does it increase, decrease, remain the same, or otherwise change?

8. Again reduce the function generator input voltage to zero and monitor the collector output. Begin increasing the input voltage slowly and note the point where the output voltage rises to its maximum point. At that time, measure the input voltage. Explain why it is necessary for the input voltage to reach that level before the output signal appears.

9. While observing the output, continue increasing the input and note any change in the output. Measure and record the output voltage.

Compare the output voltage you measured to the supply voltage V_{cc} and explain how the output value is obtained.

10. Reduce the input from the function generator to the point where the output signal at the collector just begins to drop off its maximum value. Then insert an ammeter in series with the collector supply. Measure and record the collector current.

11. Calculate and record the input power of this class C amplifier.

12. Assume now that the class C amplifier will be a frequency multiplier. The output will be the resonant frequency of the tuned circuit. Calculate all of the input frequencies of which the resonant frequency will be a harmonic. Calculate and list all the frequencies from the second through the tenth harmonics.

13. While observing the collector output, adjust the function generator input level to just where the output voltage reaches its maximum amplitude. The output level should be approximately 10 $V_{\text{p-p}}$.

14. Now adjust the function generator input frequency to each of the input frequencies you calculated above. Tune the generator carefully until the output signal peaks, and then measure the input frequency with the frequency counter. Repeat that process for all of the harmonic values you calculated in step 12.

15. As you continue to lower the input frequency and produce higher and higher harmonics, note the effect on the output voltage. Does the output increase or decrease as the harmonic number increases?

REVIEW QUESTONS

Read the question and write the answer on a separate sheet of paper.

1. The type of bias used on the class C amplifier shown in Fig. 6-1 is known as _____.
   a. Self-bias
   b. External bias
   c. Signal bias
   d. Variable bias

2. If the transistor in the class C amplifier were a PNP, the base bias with respect to ground would be _____.
   a. Positive
   b. Negative
   c. Zero
   d. Insufficient information is given to answer this question

3. In a properly adjusted class C amplifier, the collector output voltage is _____.
   a. V_{cc}
   b. $2V_{cc}$
   c. $V_{cc}/2$
   d. A value that depends upon the input signal amplitude

4. As the frequency multiplication factor increases, the output signal voltage and power _____.
   a. Decrease
   b. Increase

5. The class C amplifier multiplier deliberately distorts the signal to generate _____.
   a. Higher output voltage
   b. Higher power
   c. A clean sine wave
   d. Harmonics

ACTIVITY 6-3
CONSTRUCTION PROJECT:
FM TRANSMITTER BUG

INTRODUCTION

In this project, you will construct a small single-stage FM transmitter that radiates a signal in the FM broadcast band. A small microphone is used to provide frequency modulation. With this transmitter, you can transmit voice signals to any standard FM radio. Because of the very low power of the circuit, transmission distance is limited; in general, distances up to 100 ft can be managed reliably. This small transmitter is not miniature, but it is small enough that it could be readily concealed if you wished to use it as a bug. Let your conscience be your guide.

The circuit for this transmitter is shown in Fig. 6-2. Transistor Q_1 forms an oscillator circuit; resistors R_2, R_3, and R_4 bias the transistor into conduction. The frequency of oscillation is set by inductor L_1 and capacitors C_4 and C_5. The voltage divider action of C_4 and C_5 provides the feedback necessary for oscillation. The frequency of oscillation is determined primarily by the value of L_1 and C_5, although C_4 and the base-collector capacitance of the transistor also figure in determining the final oscillating frequency. C_5 is adjustable so that the oscillator can be tuned over a narrow range. The center frequency is approximately 100 MHz. The inductor L_1 is formed on the printed-circuit board on which the transmitter is to be constructed. The antenna is any short piece of wire. An antenna as short as 6 in. will work, but the greatest range is achieved with one that is 2 ft or longer.

The oscillator is modulated by a condenser (capacitor) microphone that is biased by R_1. The microphone capacitance varies with voice and produces a changing voltage at the input to C_1. The audio voltage is coupled to the base of the transistor, which causes the base and collector currents to vary. The voltage also causes the capacitance between the base and collector of the transistor to change; the capaci-

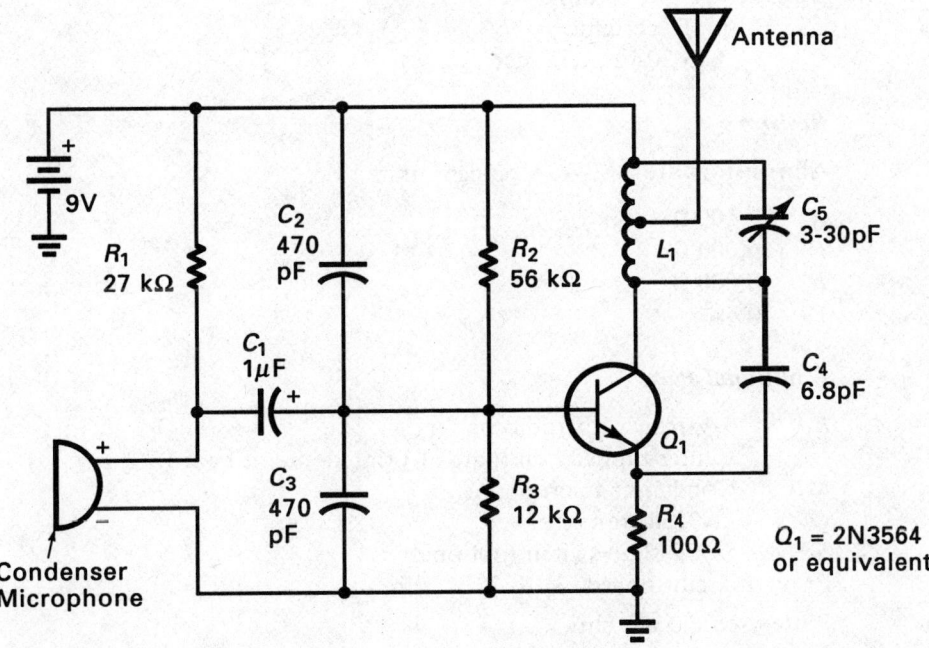

Fig. 6-2 FM transmitter bug.

53

tance exists because the junction is reverse-biased. Varying the base current varies the effective capacitance between the base and collector. Since that capacitance makes up part of the capacitance used in determining the resonant frequency, frequency modulation occurs.

CONSTRUCTION

The small transmitter is constructed on a printed-circuit board whose copper pattern is shown in Fig. 6-3. The pattern is shown actual size and, therefore, can be used as the template for making a printed-circuit board of your own. Figure 6-4 shows how to mount the components on the printed-circuit board, and the parts list follows.

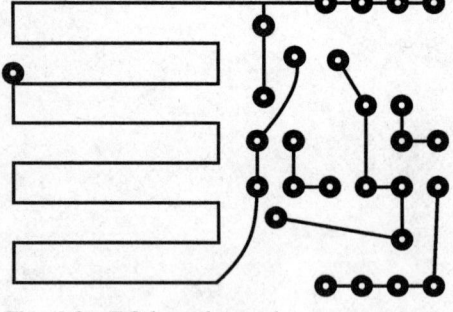

Fig. 6-3 PC board template.

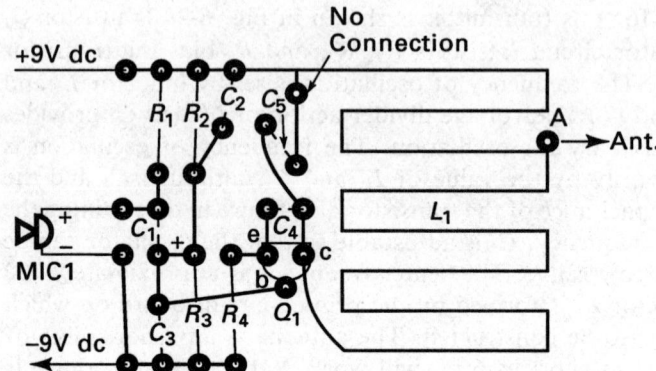

Fig. 6-4 Component placement top view—component side.

PARTS LIST

Capacitors

C_1	1-μF, 16-WVDC, electrolytic (or tantalum)
C_2, C_3	470-pF, ceramic
C_4	6-pF, ceramic
C_5	1 to 30-pF, trimmer

Resistors

All resistors are 1/2-W, 5 percent units

R_1	27,000 Ω
R_2	56,000 Ω
R_3	12,000 Ω
R_4	100 Ω

Additional Parts

B_1	9-V transistor radio battery
L_1	Center-tapped coil (part of printed-circuit board)
MIC1	Condenser microphone
Q_1	2N3564 transistor
S_1	SPST slide switch (optional)

Printed-circuit board, $3\frac{1}{4} \times 2\frac{1}{8} \times 1\frac{1}{8}$ in.

Battery connector clip

Solid hook-up wire for No. 22 antenna

Soldering equipment

TESTING

After you have constructed the circuit, test it by locating its signal on FM radio. Tune the FM radio to a blank space between stations somewhere around 100 MHz. Then connect the battery clip to the 9-V battery to turn on the transmitter. Adjust the trimmer capacitor C_5 with a non-metallic screwdriver until you hear a blanking or howl in the radio. Initial tests should be made with the transmitter right near the radio so the signal strength is high. Once you have adjusted the transmitter to the receiver frequency, make fine-tuning adjustments with the receiver tuning control. You should be able to speak into the microphone and have the sound appear at the radio speaker. To minimize the howl caused by positive feedback, move the transmitter away from the receiver. That will also give you a chance to test the maximum transmission distance.

This project was originally described in "The Dirty Little FM Snooper," by Ricky Shea, *Hands-On Electronics Magazine* (now *Popular Electronics*), pages 44–46, November 1987. You may wish to refer to the original article for additonal information. Back issues can be obtained from *Popular Electronics*.

CHAPTER 7

Communications Receivers

ACTIVITY 7-1
TEST: COMMUNICATIONS RECEIVERS

Read the question and write the answer on a separate sheet of paper.

1. The Q of a tuned circuit is proportional to _____.
   a. Resistance
   b. Inductive reactance
   c. Bandwidth
   d. Capacitance

2. To narrow the bandwidth of a tuned circuit, you must _____ the Q.

3. In a tuned circuit, $L = 2$ μH, $C = 200$ pF, $R = 0.5$ Ω, and $f = 5$ MHZ. $Q =$ _____; BW = _____.

4. A filter has 6-dB cutoff frequencies of 3100 and 3900 kHz and 60-dB cutoff frequencies of 2500 and 4500 kHz. The shape factor is _____.

5. Increasing the gain of a receiver causes its _____ to be improved.

6. To separate closely spaced signals, a receiver must have good _____.

7. The circuit that distinguishes a superheterodyne receiver from a TRF receiver is the _____.
   a. Mixer
   b. RF amplifier
   c. Demodulator
   d. AGC

8. In a superhet receiver, the incoming signal is 162 MHz. The local oscillator frequency is 117 MHz. The IF is _____.

9. A receiver has a local oscillator of 19.5 MHz and an IF of 1500 kHz. What are the signal and image frequencies? _____

10. Images may be eliminated by improving mixer _____.

11. The internal noise that has the most detrimental effect on a weak signal is produced by _____ _____.

12. A receiver has an input frequency of 18 MHz. The temperature is 80°F. The bandwidth is 500 kHz. The input resistance is 72 Ω. The noise voltage is _____.

13. To improve S/N in a receiver, the noise temperature must be _____.
    a. Increased
    b. Decreased

14. An excellent low-noise RF amplifier is made with a _____.
    a. MOSFET
    b. JFET
    c. MESFET
    d. Bipolar transistor

15. The IF amplifier in an FM receiver is used for gain and _____.

16. AGC usually controls the gain of the _____ stages in a receiver.

17. The AGC feedback voltage is _____.
    a. DC
    b. AC

18. The AGC voltage is proportional to _____.
    a. Signal frequency
    b. Deviation
    c. Signal amplitude
    d. IF gain

19. The circuit that mutes the audio in a receiver until a signal appears at the input is the _____.

20. To receive CW or code signals, a _____ _____ _____ circuit is needed in the receiver.

21. In many transceivers, a(n) _____ _____ replaces the local oscillator to improve frequency stability and precision tuning.

ACTIVITY 7-2
LAB EXPERIMENT:
NOISE AND ITS MEASUREMENT

PURPOSE

To show that noise is present in electronic circuits, its effect on low-level signals, and a way to measure noise amplitude.

MATERIALS

Qty.

1 Oscilloscope (dual trace) with direct (X1) probes
1 Function generator
1 Power supply, ±5 V dc
1 1458 dual op amp IC (individual 741 op amps can also be used)
1 MC3403 or TL074 BIFET op amp
1 0.47-μF capacitor
2 0.001-μF capacitors
2 2.2-kΩ resistors, both carbon composition and film types
2 220-kΩ resistors, both carbon composition and film types
1 1-MΩ resistor
1 10-kΩ pot

INTRODUCTION

Noise consists of random signal variations that interfere with electronic intelligence signals. It comes from a variety of sources in electronic equipment, but most of it is generated by thermal agitation in resistors and other components. Shot noise in semiconductor devices also contributes to noise. The noise itself is a random change of voltage; it varies widely in amplitude and frequency. One way to look at noise is that it contains an infinite number of frequencies mixed together in constantly changing amplitudes.

Although noise signals are low in amplitude, usually several microvolts, they can still interfere with an intelligence signal that you may want to amplify and process. Radio signals are particularly low in amplitude and, therefore, are easily interfered with by noise. In some cases, the noise is greater than the signal and completely masks it. In order for a communications system to function properly, the signal-to-noise ratio (S/N) must be high enough that the signal is intelligible. That means transmitting with sufficient power and using low-noise receivers.

The best way to ensure a good signal-to-noise ratio is to so design circuits that the noise they generate is as low as possible. Since most of the noise produced by electronic circuits is caused by thermal agitation in resistors, steps can be taken to minimize it. The noise itself is a function of the resistance at the input to the circuit, the temperature, and the bandwidth of the circuit. The larger the resistance, the higher the temperature, and the wider the bandwidth, the greater the noise that will be present. Reducing any one of the three factors will decrease the amount of noise.

Keeping input impedance low and using small values of resistance in the front end of amplifiers of electronic equipment will greatly reduce the amount of noise generated. Further, film resistors produce much lower noise voltage than the more common carbon composition resistors. Noise can be minimized by reducing the temperature and limiting the bandwidth of the circuitry. Since noise is broadband in nature, filtering it will reduce its amplitude. By restricting the bandwidth, many higher and/or lower noise frequencies are reduced or eliminated.

The types of components used in an amplifier also determine the noise. In general, JFETs and MOSFETs produce lower noise than bipolar transistors. BIFET amplifiers, those with FETs in the front end and bipolar transistors elsewhere, are good for use in low-noise applications.

In this experiment, you are going to observe actual noise in an electronic circuit by viewing it on an oscilloscope. Because noise voltages are very low in amplitude, two operational amplifiers set for high gain will be used to increase the noise level. The small amount of noise generated at the amplifier input will be multiplied by the gain of the amplifier to a level that will permit convenient viewing on an oscilloscope. You will then show how noise can be significantly reduced by using film rather than carbon resistors and by filtering. You will also show how FET amplifiers produce less noise than bipolars.

You will learn a special technique for measuring the amount of noise in a circuit. Because noise is random, it is difficult to measure and quantify. The tangential noise measurement method of using a dual-trace oscilloscope gives a surprisingly accurate measurement of rms noise voltage.

PROCEDURE

1. Construct the circuit shown in Fig. 7-1. Use the carbon composition rather than the film resistors when constructing the circuit. For the initial steps, omit the 0.001-μF capacitors from across the 220-kΩ feedback resistors. Keep all leads as short as possible to reduce 60-Hz power line noise. Adjust the 10-kΩ pot so that the input voltage is zero. You will observe the output from the second op amp at pin 7.

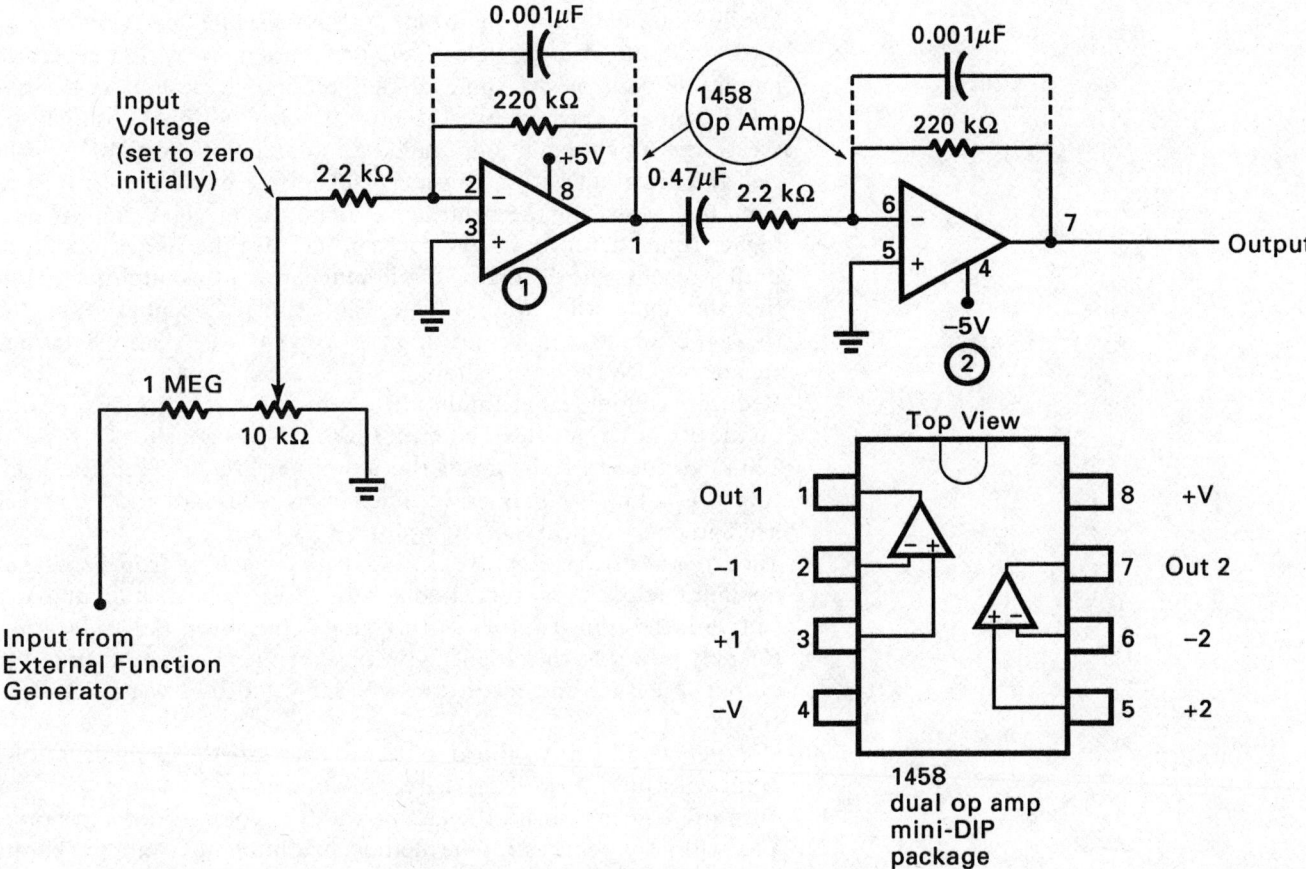

Fig. 7-1 Amplifier circuit to demonstrate noise and its measurement.

2. Knowing the gain formula of an operational amplifier inverter, calculate and record the gain of each op amp stage in Fig. 7-1 and the overall circuit gain.
3. Connect an oscilloscope to the output of the second op amp. Use a direct ($\times 1$) probe instead of the usual $\times 10$ attenuator probe. An attenuator probe contains a voltage divider which reduces the input signal by a factor of 10, and that reduction could prevent your observing the noise on the oscilloscope. Use the direct probe, and select one of the most sensitive vertical gain positions on the oscilloscope input.
4. Apply power to the circuit. Be sure that you have adjusted the positive and negative supply voltages to ± 5 V. Observe the noise at the amplifier output. Adjust the oscilloscope vertical gain and sweep controls until you observe the random varying pattern on the screen. If you have access to the intensity control on the oscilloscope, increase the trace intensity to a higher value to permit you to see more

of the noise. Because many of the noise elements have a short duration and high peaks, you may not be able to see them at the lower intensity setting. To get a more accurate noise measurement, increase the intensity so you see the major peaks more clearly. Measure the average peak-to-peak value of the noise. Since the amplitude is varying considerably, simply take an eyeball average and estimate the peak-to-peak amplitude of the noise. Record that value.

5. Compute the noise voltage level at the amplifier input by dividing the average noise output voltage by the total amplifier gain.

6. Apply a 1-kHz signal from the external function generator to the amplifier input. Adjust the 10-kΩ pot slowly while observing the amplifier output. You may also have to adjust the function generator amplitude control. At some point, you should be able to see the 1-kHz signal appear above the noise level. Continue to slowly increase the input signal amplitude. At some point the signal will be clearly visible but with a considerable amount of noise riding on it. Continue increasing the input signal amplitude until it is significantly higher than the noise itself. You can increase the signal amplitude until it reaches the distortion or clipping level of the amplifiers. Note that the higher the input voltage, the cleaner the signal looks and the less effect that noise has upon it. You can easily see why a high signal-to-noise ratio is desirable.

7. Reduce the input signal amplitude to zero and again display the noise on the oscilloscope. Next, connect the 0.001-μF capacitors across the 220-kΩ feedback resistors in the amplifier circuit. Keep the leads short to minimize 60-Hz power line pickup. Measure and record the average peak-to-peak noise amplitude.

8. Turn off the power. Remove the 0.001-μF capacitors from across the op amp feedback resistors. Replace the resistors in each op amp circuit with the film resistors of the same value. Keep the leads short.

9. Reapply power to the circuit. Now observe the noise at the amplifier output. Measure and record the average peak-to-peak noise output voltage.

10. Use the overall gain of the amplifier to calculate the equivalent noise input with the film resistors. Record your answer.

11. Turn off power. Rewire the circuit with the composition resistors.

12. You will now demonstrate a method of obtaining a more accurate measurement of actual noise voltage by using a calibrated dual-trace oscilloscope. The oscilloscope should have an alternate horizontal sweep capability. Connect the direct probes from the two vertical input channels to the amplifier output at pin 7. The noise will be displayed simultaneously on both channels. Adjust the vertical calibrations to exactly the same settings so that the noise on each channel is displayed at the same level.

13. Next, by using the vertical-position controls on one or both vertical-input channels, adjust the traces so that they begin to merge. Your goal is to adjust the noise waveforms so that the darker background area between the two traces just disappears.

14. Next, remove the oscilloscope connections to the amplifier output and thereby reduce the input voltage to zero on both channels. At this time the oscilloscope display will simply show two horizontal traces separated by a small distance that is a measure of the noise amplitude. Use the calibrated scope graticule to determine it. The voltage that you measure between traces is 2 times the actual rms noise voltage. Record the actual noise voltage.

15. As before, divide the value in step 14 by the total amplifier gain to obtain the actual input noise level. Record the result.

16. Turn off power and rewire the circuit with the MC3403 or TL074 BIFET op amps. Refer to the data sheets to obtain pin connections, because they are different from those shown in Fig. 7-1.
17. Apply power to the circuit. Again measure the output noise and then calculate the input noise. Use the tangential noise measurement technique. Record your results. Is the noise produced by the BIFET amplifiers more or less?
18. Turn off power and disconnect the circuit.

REVIEW QUESTIONS

Read the question and write the answer on a separate sheet of paper.

1. Which of the following items does not directly affect the noise voltage in an electronic circuit? _____
   a. Bandwidth
   b. Temperature
   c. Resistance
   d. Signal frequency

2. The $0.001\text{-}\mu\text{F}$ capacitors added to the circuit cause which component of the noise to be greatly minimized? _____
   a. High frequency
   b. Low frequency
   c. Midrange frequencies
   d. 60-Hz power line noise

3. (True or False) Carbon composition resistors generate more noise than film resistors. _____

4. (True or False) Bipolar transistors produce less noise than FETs.
   _____

5. An amplifier has an input resistance of 75 Ω, a total bandwidth of 36 MHz, and temperature of 30°C. The noise voltage is _____ μV.
   a. 3.11
   b. 4.5
   c. 6.72
   d. 8.09

ACTIVITY 7-3
LAB EXPERIMENT:
FREQUENCY SYNTHESIZER

PURPOSE

To demonstrate the operation and characteristics of a phase-locked loop frequency synthesizer.

MATERIALS

Qty.
1   Oscilloscope
1   Function generator
1   Frequency counter
1   Power supply (± 12 V)
1   565 PLL IC
2   4029 CMOS binary/BCD counters
1   $0.001\text{-}\mu\text{F}$ capacitor
(*continued*)

1 0.01-μF capacitor
1 0.1-μF capacitor
1 0.47-μF capacitor
2 560-Ω resistors
1 2.2-kΩ resistor
1 3.3-kΩ resistor

INTRODUCTION

A frequency synthesizer is a signal generator circuit producing a highly stable output frequency that can be changed in discrete increments. A phase-locked loop can be used as a frequency synthesizer by connecting a stable crystal oscillator as a reference to the PLL input and taking the output from the VCO; see Fig. 7-2. By connecting a variable-frequency divider between the frequency output and the phase detector input, the output frequency can be changed by simply changing the frequency division ratio.

There are two key factors in the design of the phase-locked loop frequency synthesizer. First, the output frequency of the synthesizer is equal to the reference input frequency times the frequency division ratio N.

$$f_o = Nf_r$$

As you can see, changing the frequency division ratio will vary the output frequency. One way to vary the divide ratio is to use a preset down counter. A binary number is loaded into the counter, which is then decremented to zero. Then the cycle is repeated. For example, if the binary number 0110 is loaded, the counter will divide by 6. A variety of variable or programmable IC counters are available to perform division in a PLL synthesizer. In this experiment, you will use the 4029 4-bit up-down counter, which can be set for binary or BCD counting.

Second, the frequency synthesizer output frequency varies in increments equal to the reference frequency. For example, if the reference frequency is 100 kHz, the synthesizer output will be varied in 100-kHz

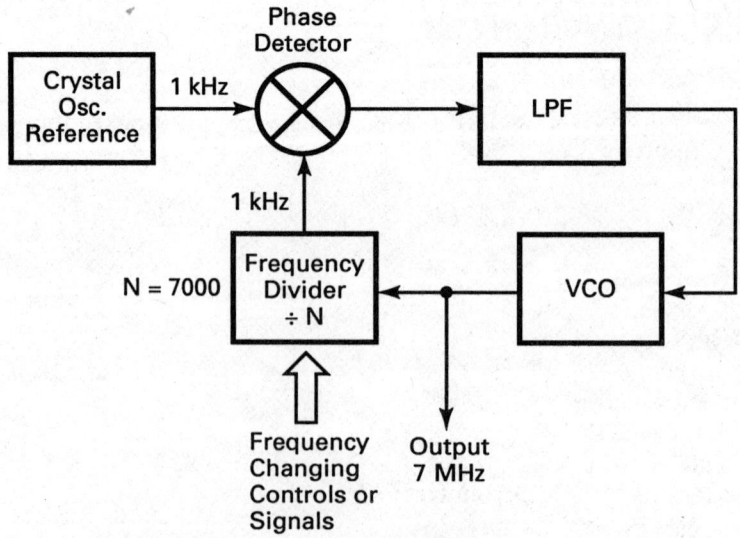

Fig. 7-2 Phase-locked loop frequency synthesizer.

steps by varying the frequency division ratio. In some PLL synthesizers, a frequency divider is used between the reference oscillator and the phase detector input to decrease the frequency step size.

In this experiment, you will construct a PLL frequency synthesizer and verify these characteristics.

PROCEDURE

1. Construct the circuit shown in Fig. 7-3 on the next page. This is the PLL circuit you used in a preceding experiment, but note the different R and C values. You will use a 600-Hz input square wave signal from an external function generator as the reference signal. Initially you will use a frequency divider made up of a 4029 CMOS counter. Pin 9 on the 4029 determines whether the count is pure binary (pin 9 at +12V) or BCD (pin 9 at ground).

2. Determine the frequency division factor of the 4029. Record the result. Check the state of pin 9 in Fig. 7-3; then assume 4-bit operation with the output taken from pin 2.

3. Examine the circuit shown in Fig. 7-3 and predict the output frequency of the VCO. Record your prediction.

4. By using a frequency counter, set the reference input from the function generator to 1 kHz. Measure and record the VCO output frequency at pin 4 on the 565, which is the synthesizer output. Are your computed and measured values the same?

5. To verify that the PLL is locked, vary the reference input frequency over a narrow range and observe the VCO output. What happens?

6. Change the circuit by moving pin 9 on the 4029 from +12V to ground. The 4029 now divides by what?

7. Set the reference input to 1 kHz. Predict the VCO output. Record your prediction.

8. Measure and record the VCO output frequency at pin 4 on the 565.

9. Rewire the circuit so that it is as shown in Fig. 7-4 on page 65. Change the 3.3-kΩ resistor on the 565 between pins 8 and 10 to 2.2-kΩ. Two 4029 ICs are cascaded between the VCO output and phase detector input. Looking at the state of pin 9 on each 4029 determines the individual IC divide ratios and the combined or total divide ratio. What is it? Predict the synthesizer output frequency if the function generator (reference) is set to 100 Hz. Record your prediction.

10. Set the function generator to 10 Hz and apply power. Measure and record the output frequency of the VCO with a frequency counter. Are your computed and measured values the same? If the divide ratio of the 4029 IC were variable, in what steps would the synthesizer frequency change?

11. Rewire the circuit so that it is as shown in Fig. 7-5 on page 66. One 4029 is used in the PLL feedback loop, and the other is used between the function generator and the reference input to pin 2 on the 565. Predict the VCO output frequency if the function generator is set to 10 kHz. Record your prediction.

12. Determine and record the incremental amount by which the frequency will vary as the frequency division ratio is changed.

13. Set the function generator to 10 kHz. Measure and record the VCO output frequency. Compare it to the predicted value. Are your measured and calculated values the same?

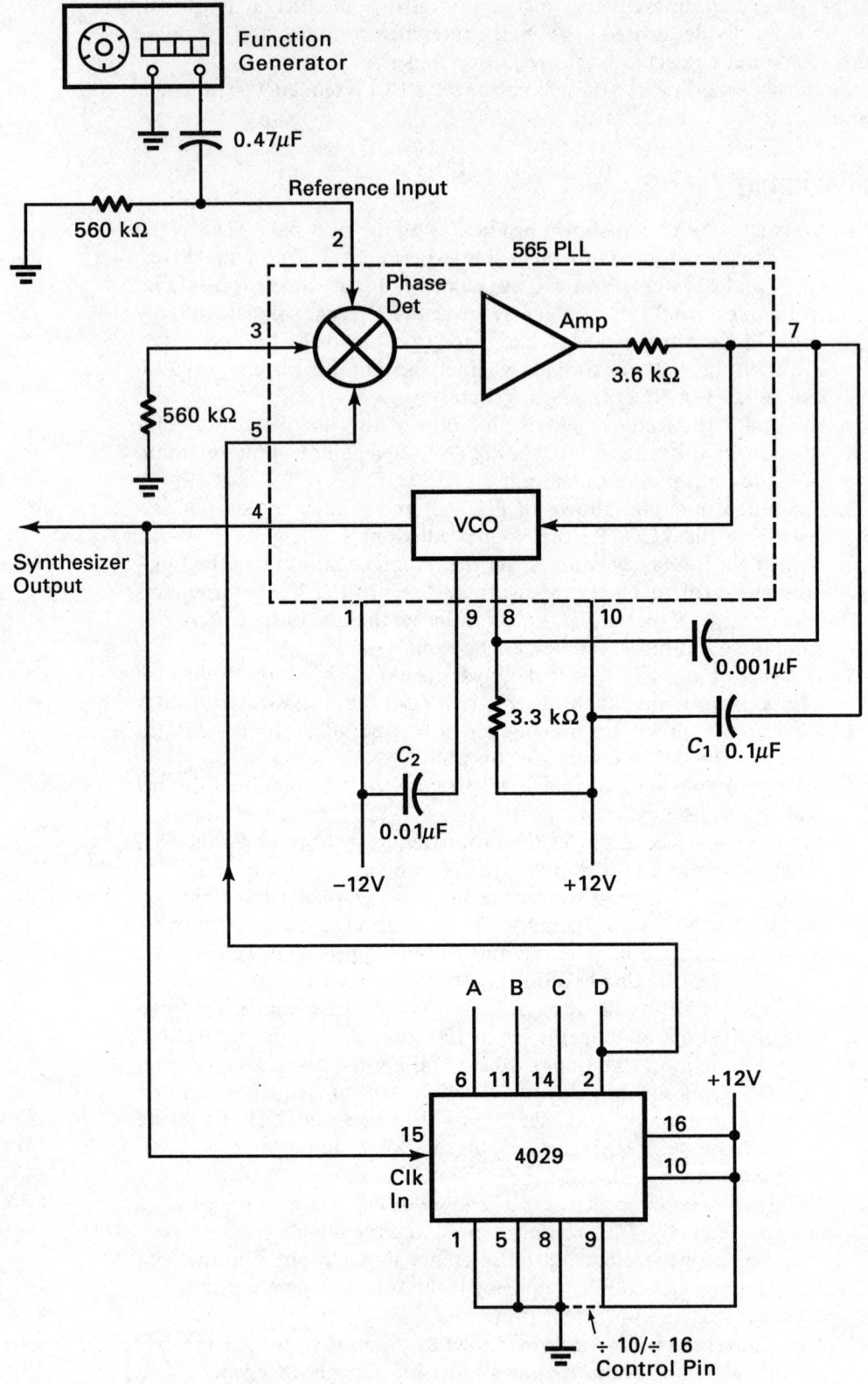

Fig. 7-3 PLL synthesizer experimental circuit.

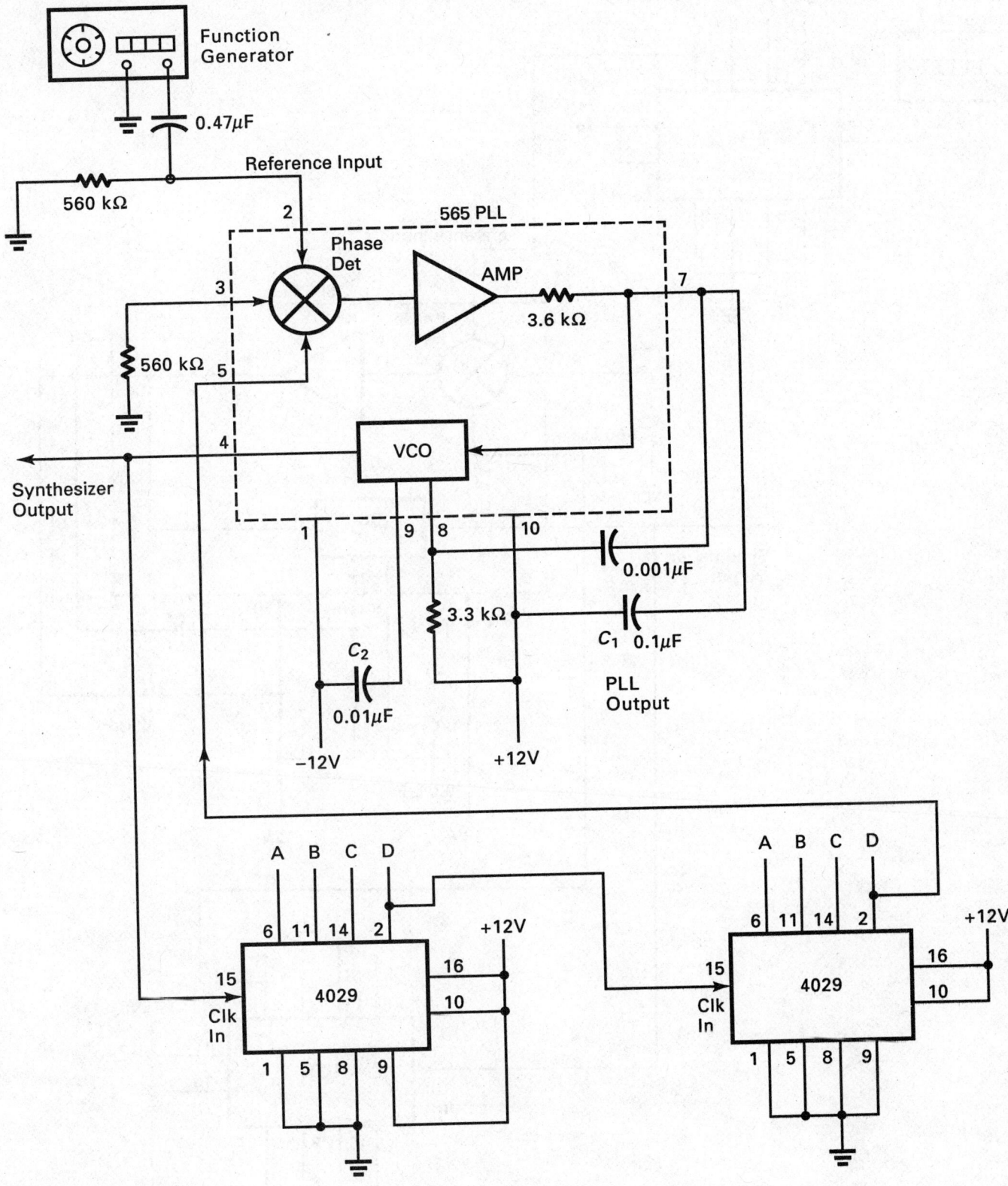

Fig. 7-4 Revised synthesizer experimental circuit.

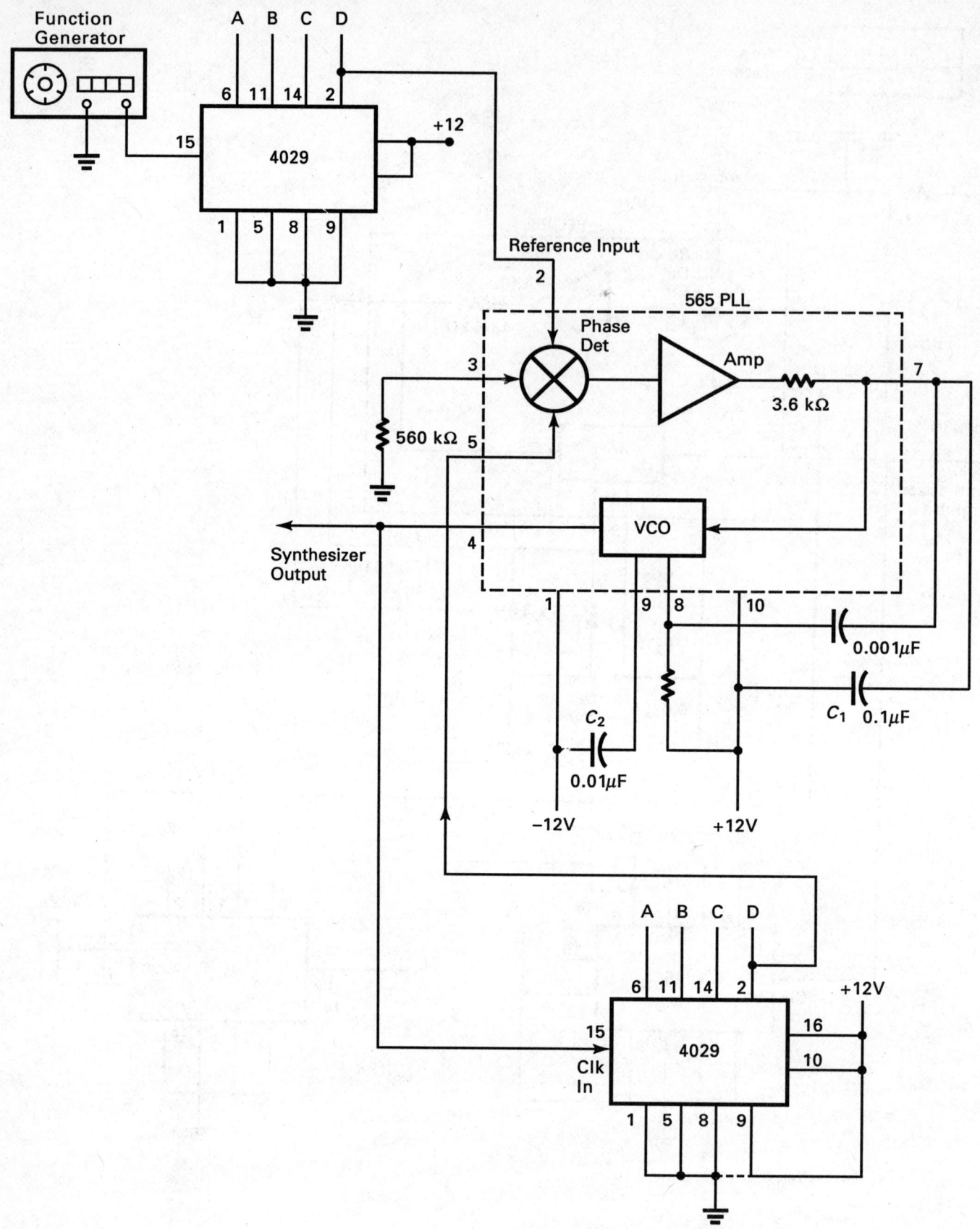

Fig. 7-5 PLL synthesizer with dividers at the reference input and in the feedback loop.

REVIEW QUESTIONS

Read the question and write the answer on a separate sheet of paper.

1. The frequency variation increment of a PLL synthesizer is determined by the _____.
   a. VCO free-running frequency
   b. Frequency divider ratio
   c. Both *a* and *b*
   d. Reference frequency

2. A PLL synthesizer has a 1-MHz reference and a frequency divider of three cascaded BCD counters. The output frequency is _____.
   a.   1 MHz
   b. 10 MHz
   c. 100 MHz
   d.   1 GHz

3. In the circuit described in question 2, a two-decade BCD counter is connected between the reference and the PLL phase detector input. The output frequency is _____.
   a. 10 kHz
   b. 100 kHz
   c. 10 MHz
   d. 100 MHz

4. (True or False) The PLL synthesizer can function as a frequency multiplier. _____.

5. To function properly, the VCO in the PLL must have its free-running frequency set approximately to the _____.
   a. Desired output frequency
   b. Output divided by *N*
   c. Frequence division ratio
   d. Reference frequency

ACTIVITY 7-4
CONSTRUCTION PROJECT: BUILD A RECEIVER

PURPOSE

To build a complete superheterodyne receiver from a commercial kit and experiment with it.

MATERIALS

Oscilloscope

Function generator

Digital multimeter

Receiver kit as supplied or recommended by the instructor

PROCEDURE

1. Open the receiver kit and read the instruction manual through once quickly to familiarize yourself with the project.
2. Obtain all of the necessary tools (soldering iron, needle-nose pliers, side cutters, screwdrivers, etc.) and solder.

3. Build the receiver by carefully following the instructions given in the kit manual.
4. Perform the individual experiments given in the kit manual. Follow the instructions given. Some kits will have you build the receiver one circuit at a time and then perform related experiments rather than build the entire receiver first.

CHAPTER | 8

Multiplexing

ACTIVITY 8-1
TEST: MULTIPLEXING

Read the question and write the answer on a separate sheet of paper.

1. The process of transmitting two or more signals over a common communications medium by sharing a fixed spectrum space is known as _____.

2. The process of separating multiple signals sharing a common channel is called _____.

3. (True or False) Any type of modulation can be used with FDM.
_____

4. Circuits that demultiplex signals sharing a common frequency band are called _____.

5. A popular modulator circuit used in FM/FM multiplexing systems is the _____.

6. The center frequency of each channel in an FDM system is called a(n) _____.

7. In telephone FDM systems, the type of modulation used is _____.

8. The type of modulation used for the *LR* channel in FM stereo multiplex systems is _____.

9. Time division multiplexing uses _____ modulation.

10. An analog signal with a maximum frequency content of 4500 Hz must be sampled at a minimun rate of _____ kHz to maintain intelligibility.

11. A component widely used for sampling in a TDM circuit is the _____.

12. (True or False) FDM and TDM techniques may not be mixed.
_____

13. Another name for TDM with serial digital signals is _____.

14. Compressing the amplitude range of an analog signal to overcome noise and quantization error is known as _____.

15. A popular PCM format used in digital telephone systems has the designation _____.

ACTIVITY 8-2
LAB EXPERIMENT:
PULSE AMPLITUDE MODULATION
AND TIME DIVISION
MULTIPLEXING

PURPOSES

To construct a circuit that will permit demonstration of both pulse amplitude modulation and time division multiplexing.

MATERIALS

Qty.

1 Oscilloscope
1 Sine wave signal source
1 Square wave signal source
1 555 timer IC
1 4016 or 4066 quad bilateral CMOS switch IC
1 NPN transistor (2N3904, 2N4401, etc.)
2 10-kΩ pots
1 1-kΩ resistor
1 4.7-kΩ resistor
2 15-kΩ resistors
1 22-kΩ resistor
1 0.01-μF capacitor
1 10-μF electrolytic capacitor

INTRODUCTION

In pulse amplitude modulation, the information signal to be transmitted is turned off and on periodically by a gate circuit. The gate, usually some kind of switch, allows segments of the intelligence signal to be passed through. The process is generally known as sampling the intelligence signal. The output of the gate is a series of periodic pulses whose amplitude follows the intelligence signal. This series of pulses is then used to modulate a carrier.

A key design factor in a PAM modulator circuit is the rate of sampling. To represent the intelligence signal accurately, the sampling rate must be at least 2 times the highest frequency content contained in the modulating signal. Higher sampling rates of 10 or more samples for the highest frequency in the intelligence signal is more desirable because it more accurately represents the information being transmitted.

The concept of PAM can be expanded to produce time division multiplexing. A time division multiplexer has two or more inputs or channels, each of which is sampled at a high rate of speed. A gate on each input intelligence signal samples its respective signal for a short period of time. Each of the signals is sampled in sequence one at a time so that a segment of each signal appears in the output. For example, a four-input time division multiplexer would sequentially sample four input signals one at a time one after the other. The cycling sequence then repeats again and again. The resulting time division–multiplexed output signal is a composite of the four intelligence signals sampled sequentially. The result is that four intelligence signals can be transmitted simultaneously over a single channel.

In this experiment you are going to demonstrate PAM and time division multiplexing. You will use a solid-state switch that is operated by an astable multivibrator. You will observe both PAM and time division signals at the output.

The gate or switch used in this experiment is a CMOS switch. It is made of both P- and N-channel enhancement mode MOSFETs connected in such a way that they act as a simple single-pole single-throw switch. When a binary 1 logic signal is applied to the switch, the switch turns on. When a logic signal is binary 0, the switch turns off. During the on period, the intelligence signal is passed. The 4016 and 4066 ICs contain four sampling switches.

One of these switches will be used to demonstrate PAM, and two of them will be connected together to form a two-channel multiplexer. A 555 IC timer connected as an astable multivibrator provides the sampling signal.

PROCEDURE

1. Construct the circuit shown in Fig. 8-1. The 555 IC is used as an astable multivibrator for generating a square wave that will operate

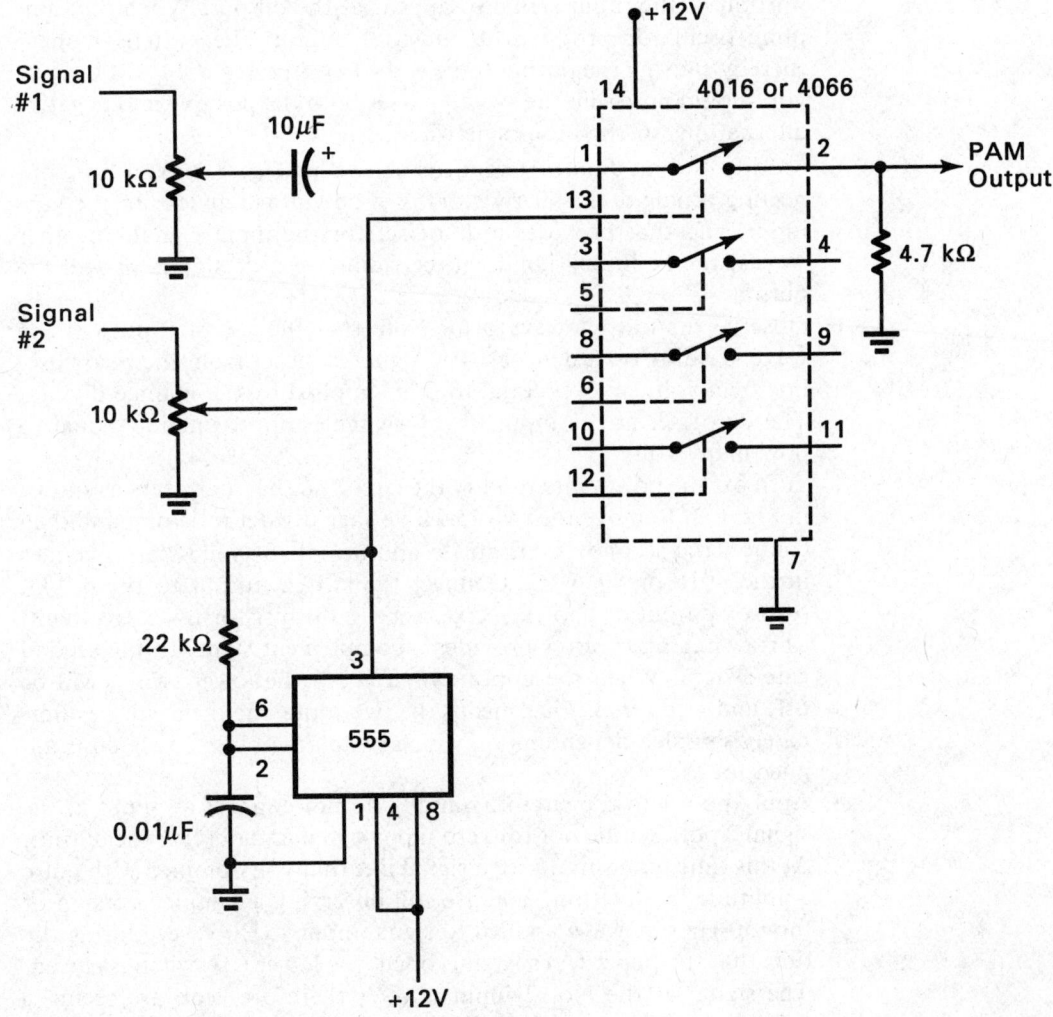

Fig. 8-1 PAM modulator.

the CMOS switch. The resistor and capacitor associated with the 555 sets the 555's oscillating frequency. The 4016 or 4066 IC contains four CMOS switches, and you will use one of them as a gate to produce a PAM modulator circuit. Your signal source can be a function generator or any other source of a sine wave intelligence signal. Apply the sine wave to the signal 1 input. The 10-kΩ pot will be used to adjust the signal amplitude. The output is taken from across the 4.7-kΩ resistor at the switch output. Check to be sure that each integrated circuit is connected to +12 V and ground before beginning.

2. Set the function generator supplying signal 1 for 150 Hz with an amplitude of several volts peak to peak. Then while monitoring the signal at the arm of the 10-kΩ potentiometer, adjust the pot for a signal amplitude of approximately 0.5 V_{p-p}.

3. Apply power to the circuit. With an oscilloscope, observe the 555 oscillator output at pin 3. Use the calibrated horizontal sweep on the oscilloscope or a frequency counter to determine the frequency of the sampling signal. Record your result.

   Considering the frequency of the input intelligence frequency and the sampling frequency above, is the intelligence signal sampled fast enough to retain intelligibility?

4. Connect the oscilloscope to the 4.7-kΩ output resistor. You should observe a PAM output signal. Each time the sampling signal is a binary 1, or approximately +12 V, the switch closes. That allows a portion of the input signal to appear at the output. When the sampling oscillator produces a binary 0 output, the switch is open, thereby causing the output to be zero. In observing the PAM output, you should note that the switch passes both the positive and negative alternations of the input sine wave.

5. Modify the modulator circuit as shown in Fig. 8-2. You are connecting a pair of 15-kΩ resistors as a dc voltage divider to the gate input, and that provides a dc offset for the input signal. In other words, the ac input signal will be riding on a dc signal, as will the output.

6. Observe the output waveform. Note that the sampled output sine wave is riding on a dc level. The signal is offset from the zero baseline by a voltage level equal to that supplied by the voltage divider. The dc offset at the input to the switch is approximately equal to how many volts?

7. Turn off the power and modify the circuit so that it appears as shown in Fig. 8-3. Remove the two 15-kΩ voltage divider resistors, and then connect the second 10-kΩ pot to another of the soild-state switches in the 4016 or 4066 IC. Connect the switch output to the 4.7-kΩ resistor output as shown. Next, wire the transistor inverter circuit. This switching inverter provides a complement signal to the second gate switch. When the upper switch is on, the lower switch will be off, and vice versa. That means the two input channels will be alternately sampled depending on which switch the binary 1 signal is applied to.

8. Apply power to the circuit. Initially do not connect an input at the signal 2 pot. Set the pot for zero input signal and observe the output. At this time, you should see a signal like that you obtained with pulse amplitude modulation. You are still observing a sampled version of the 150-Hz sine wave applied to signal input 1. However, during the time that the upper (A) switch is open, the lower (B) switch is closed. The signal at the No. 2 input is 0 V; therefore, you are seeing a 0-V level during the time that the B switch is closed. Applying

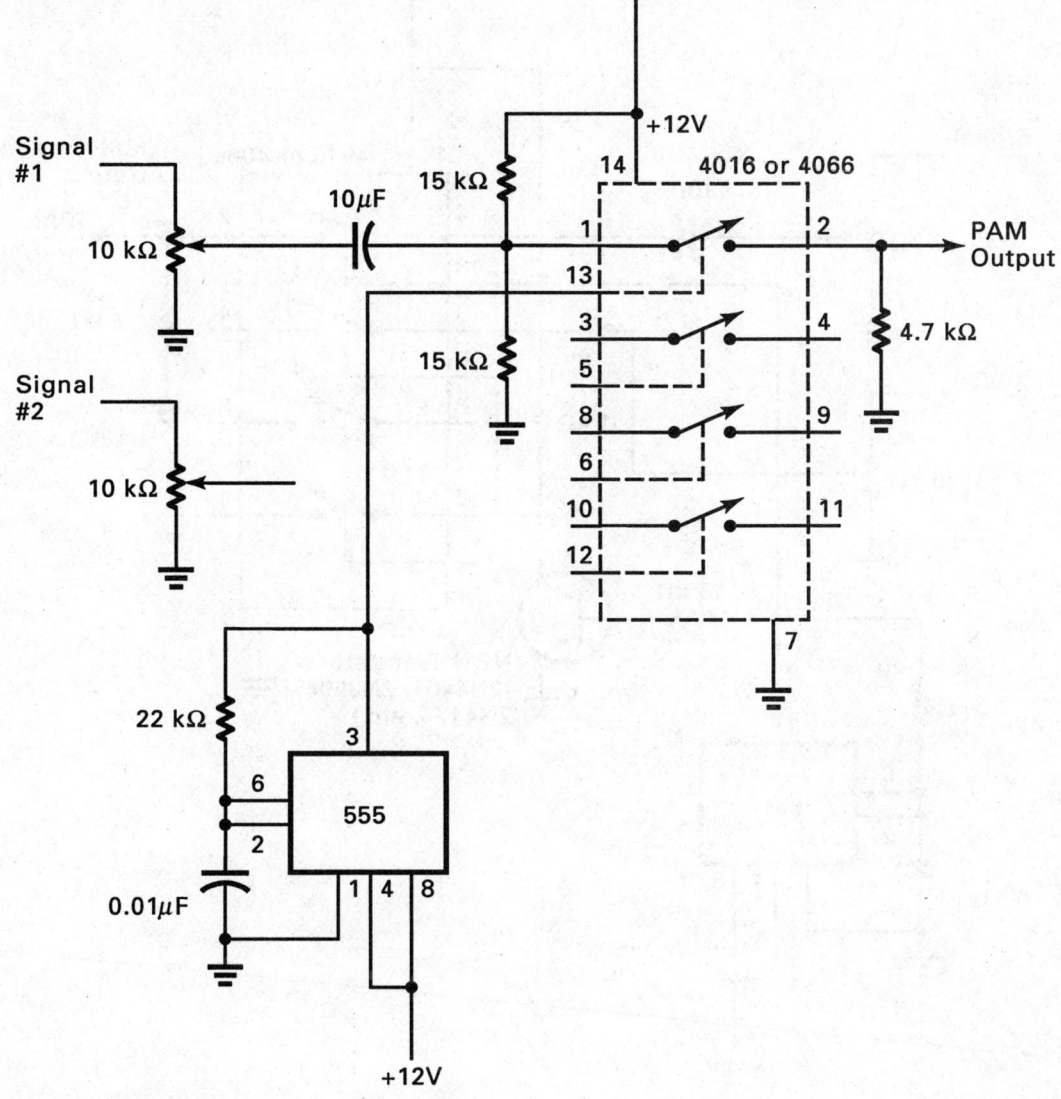

Fig. 8-2 PAM modulator with dc offset.

another signal to the No. 2 input will cause that signal to appear at the output during the time the B switch is closed.

9. Connect a positive dc voltage to the signal No. 2 input. Adjust the dc voltage at the arm of the 10-kΩ pot for a value of approximately 0.25 to 0.3 V. Now observe the multiplexed output. What you are observing is the sampled 150-Hz sine wave and a dc voltage level other than zero. While observing the output signal, adjust the 10-kΩ pot on input signal 2 so you can see the output variation. You can take this one step further and reverse the polarity of input signal 2 to show the effect.

10. Remove the dc voltage from the signal 2 input. Using a second function generator, connect a square wave to the signal 2 input. Set the frequency for approximately 200 Hz and an output that varies between zero and +5 V.

11. Observe the multiplexed output waveform. With all of the different signals involved—the 150-Hz sine wave, 200-Hz square wave, and the sampling oscillator signal—your oscilloscope will go crazy. By

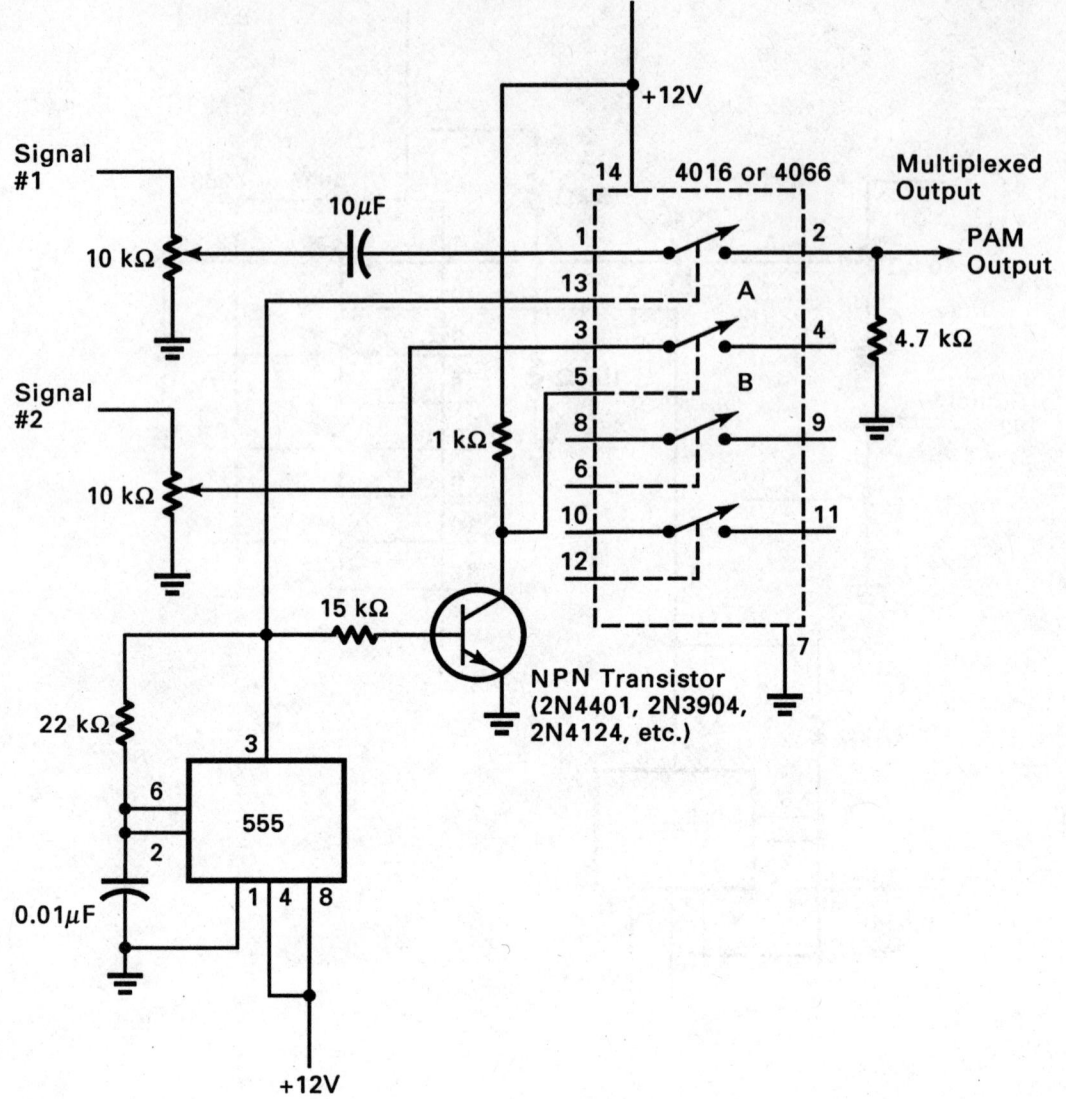

Fig. 8-3 Two-channel time division multiplexer.

using the horizontal frequency and the triggering controls, you should be able to obtain a stable output waveform. Making minor adjustments in the square wave input frequency will also help to stabilize the waveform for viewing. Of course, what you should see is a sine wave and a square wave appearing in the output, but sampled. The samples are, of course, interlaced, which causes the signals to be merged. If additional input signals were connected, these too would be merged with the composite signal. The resulting output signal is relatively complex and meaningless by itself. However, when it is applied to a demultiplexer circuit, the different signals will be unscrambled and the original modulating signals will be recovered intact.

CHAPTER | 9

Antennas and Transmission Lines

ACTIVITY 9-1
TEST: ANTENNAS AND
TRANSMISSION LINES

Read the question and write the answer on a separate sheet of paper.

1. Name the two basic types of transmission lines. *twisted pair coax*

2. The wavelength of an 18-MHz signal is _____ m.

3. A 25-m signal has a frequency of _____ MHz.

4. One wavelength of a 250-MHz signal is _____ ft.

5. The characteristic impedance of a transmission line is dependent upon the distributed *cap* and *ind* of the line.

6. A coax line has a center conductor with a 0.125-in. diameter and an outer shield with a 0.40-in. inside diameter. The characteristic impedance is _____ Ω.

7. Match the transmission line type to the characteristic.
   - *b* Twin lead   *a.* balanced
   - *a* Coax   *b.* unbalanced

8. The impedance of twin lead is _____ Ω.

9. Two common coax characteristic impedances are approximately *50* and *75* Ω.

10. A coax line has an attenuation of 8 dB per 100 ft. A run of 175 ft has an attenuation of *14* dB.

11. (True of False) That a transmission line has reflected power means that the line is terminated in its characteristic impedance. *F*

12. The phenomenon that occurs when a transmission line is not terminated in its characteristic impedance is called *REFLECTION*

13. SWR produced by a 50-Ω transmission line and a 73-Ω load is _____.

14. Sinusoidal distributions of voltage and current along a transmission line not terminated in its characteristic impedance are known as _____.

15. The optimum SWR value is *1*.

16. Any SWR less than _____ is usually acceptable in most applications.

17. Shorted and/or open one-quarter and/or one-half wavelength lines act like _____.

18. The SWR value tells how much _____ is lost in a line.

19. A shorted quarter wave line acts like a _____.
    a. Series resonant circuit
    b. Parallel resonant circuit

20. An open half-wavelength line acts like a _____.
    a. Series resonant circuit
    b. Parallel resonant circuit

21. The speed of a signal in a transmission line compared to that in free space is __A__.
    a. Less
    b. More

22. One-half wavelength of coax with a velocity factor of 0.62 at 350 MHz is _____ in.

23. An electric field combined perpendicular to a magnetic field produces a(n) electro magnetic

24. An antenna has a magnetic field that is horizontal to the earth. Its polarization is __b__.
    a. Horizontal
    b. Vertical

25. The impedance of a half-wave dipole is __73__ Ω.

26. A half-wave dipole has a length of _____ in. at 220 MHz.

27. The greatest signal strength is received or transmitted by a dipole when it is positioned how with respect to the signal source or destination? __b__
    a. End to end
    b. Broadside   right angles

28. Most vertical antennas have which type of antenna pattern? _____
    a. Omnidirectional
    b. Bidirectional
    c. Unidirectional
    d. Quasidirectional

29. You see a quarter-wave vertical with a length of 8 in. Its operating frequency is _____ MHz.

30. The names of the elements of a Yagi antenna are driven, reflector, and director.

31. Which element of a Yagi do you point at a signal source for maximum signal? _____
    a. Reflector
    b. Director
    c. Driven element
    d. Boom

32. A measure of antenna directivity is _____.

33. Antennas with high directivity also have high _____.

34. Two popular types of driven array gain antennas are the _____ and _____.

35. A popular antenna with variable-length driven elements is the _____.

36. The ionosphere causes radio signals to be ___b___.
   a. Reflected
   b. Refracted

37. The ionosphere causes radio signals below about _20_ MHz to be bent back to earth.

38. (True of False) The ionosphere has little effect on microwave signals.
   ___T___

39. At VHF, UHF, and microwave frequencies, transmission is primarily by ___c___.
   a. Ground wave
   b. Sky wave
   c. Direct wave

40. The line-of-sight distance between a transmitting antenna 30 ft high and a receiving antenna 65 ft high is _9.15_ mi.

41. A repeater station consists of a(n) _____ and a(n) _____ operating on different frequencies.

ACTIVITY 9-2
LAB EXPERIMENT:
RADIO ANTENNA

PURPOSE

To construct a dipole antenna and test its directional characteristics.

MATERIALS

FM radio with terminals for attaching an external antenna

300-Ω twin lead (about 8 to 15 ft)

Soldering iron, solder, side cutters, and knife or wire strippers

PROCEDURE

1. Use twin lead to construct a folded dipole antenna for your favorite FM radio station, whose frequency f you must know. What is it?
2. Cut the antenna following the guidelines in the text. It must be one-half wavelength long, but take into account the velocity factor of twin lead, which is 0.8. What length do you find?
3. Strip both ends of the twin lead and twist the leads together, as Fig. 9-1 on the next page shows. Cut into the center of one conductor of the twin lead and attach the 300-Ω transmission line; see Fig. 9-1. The transmission line length is not critical, but it should be great enough to attach the line to the receiver and provide for movement of the antenna. A minimum length is about 5 ft, but greater is better.
4. Connect the transmission line to antenna terminals on the FM receiver. Turn on the receiver and tune it for your station.
5. Hold the antenna in your hands with your arms extended so the antenna is stretched out horizontally to its full length. Stand away from the receiver and any obstructions (wall, furniture, etc.) as far as possible. To determine the directional characteristic of the antenna, rotate your body slowly 360° while monitoring the signal strength. Note how the signal amplitude goes up and down as you turn. Predict the antenna directivity pattern from your observations.

Note: If you do not get clear, pronounced variations in signal

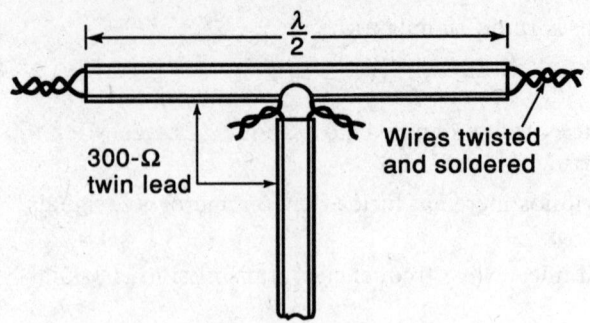

Construction with twin lead

Fig. 9-1 Folded dipole construction with twin lead.

strength, try a different position in the room. Some obstruction may be interfering with normal operation. Also, try another but weaker station. Strong signals obscure the effect. Tune to the weakest signal you can find and then repeat this step.

Can you determine the direction of the transmitting station from your prediction?

6. Repeat step 5 on another station of a different frequency. Does the antenna work well on this frequency?

7. Hold the antenna vertically and repeat the procedure that is described in step 5. What is the shape of the directivity pattern?

CHAPTER | 10

Microwave Techniques

ACTIVITY 10-1
TEST: MICROWAVE TECHNIQUES

Read the question and write the answer on a separate sheet of paper.

1. Signals whose frequencies are above 1 GHz are known as _____.

2. Two common applications of microwave signals are _____ and _____ communications.

3. Because the microwave region offers extremely broad spectrum space, it is suitable for wideband signals such as _____, and _____.

4. (True or False) Conventional electronic components are widely used in microwave circuits. _____

5. (True of False) Coax transmission line cannot be used for microwaves. _____

6. The most common microwave transmission line is called _____.

7. Printed-circuit board techniques are used to make _____ and _____, which are used to form a short transmission line at microwave frequencies.

8. A waveguide acts like a _____ filter.

9. The cutoff frequency of a waveguide 4 by 2 cm is _____ GHz.

10. A microwave signal is introduced into a waveguide by either a(n) _____ or a(n) _____.

11. The two types of field patterns setup in a waveguide are designated _____ and _____.

12. A cavity resonator can be made from a shorted half-wave section of _____.

13. A cavity resonator performs like a _____.
    a. Parallel tuned circuit
    b. Series tuned circuit

14. If the size of a cavity is changed, its _____ varies.

15. Cavities have _____.
    a. Low Q
    b. High Q

16. Name eight types of semiconductor diodes commonly used in microwave circuits and one typical application of each. _____

17. A high-power vacuum tube using cavities for microwave amplification is the _____.

18. A reflex klystron is used as a(n) _____.

19. A magnetron is a microwave tube used as a(n) _____.
    a. Oscillator
    b. Amplifier
    c. Mixer
    d. Demodulator

20. The most widely used microwave power amplifier tube is the _____ _____ _____.

21. Wide-bandwidth microwave power amplification is best obtained with a _____.
    a. Klystron
    b. TWT
    c. Transistor

22. Narrowing the beam width of a horn antenna causes its gain to _____.
    a. Decrease
    b. Increase

23. The power gain of a pyramidal horn antenna at a frequency of 6 GHz with an aperture A of 0.02 m² is _____ dB. Assume K = 0.5.

24. A parabolic reflector causes the gain of the antenna used with it to _____.
    a. Increase
    b. Decrease
    c. Remain the same

25. The beam width of a parabolic reflector antenna with a diameter of 2.5 m at 15 GHz is _____ degrees.

26. The gain of the antenna described in question 25 is _____ dB.

27. (True of False) A horn antenna may have either vertical or horizontal polarization. _____

28. A Cassegrain feed parabolic antenna is one with a(n) _____ at the focal point.

29. Helical antennas produce which type of polarization? _____
    a. Vertical
    b. Horizontal
    c. Circular

30. A microwave antenna made of circular waveguide called the _____ has a(n) _____ radiation pattern.

ACTIVITY 10-2
PROJECT: RADAR DETECTOR

PURPOSE

To examine a commonly available microwave receiver to determine its contents and construction.

MATERIALS

Any modern police radar detector receiver.

INTRODUCTION

The popular and widely used radar detectors (fuzz busters) are really state-of-the-art microwave receivers. Most of these units are capable of receiving both X-band (10.525-GHz) and K-band (24.15-GHz) signals from police Doppler radar guns. The receivers are superheterodyne units with both audible and visible readout indicators that signal the presence of a radar signal. These receivers have a sensitivity that allows them to pick up signals up to 3 mi away depending upon the radar frequency, the terrain, and other factors.

In this experiment, you will examine one of these receivers to determine what components are used in it and how the unit is constructed.

PROCEDURES

1. Open the receiver cabinet by using whatever tools may be necessary. Be careful not to break anything, and remember to save any screws, nuts, or other small parts that must be removed to gain access to the unit.
2. Examine the receiver. What type of receiving antenna is used? Is its gain higher or lower on the K or X band?
3. What types of tuned circuits are used?
4. What kind of mixer is used?
5. What type of local oscillator is used?
6. What might be the frequency of the local oscillator given the fact that two input signals (X and K bands) must be detected?
7. What type of signal indicators or signal strength readouts are used?
8. What kind of demodulator is used?

ACTIVITY 10-3
PROJECT: MICROWAVE OVEN

PURPOSES

To examine a microwave oven to become familiar with its construction and operation.

MATERIALS

Any common microwave oven
Service manual for microwave oven
Neon bulb

INTRODUCTION

In addition to its use in communications, microwave energy is used for heating. The popular microwave oven is an example. A magnetron tube is used as a microwave oscillator to produce a high-frequency signal that will heat liquids and food. The signal from the magnetron cavities is coupled by a waveguide assembly to the cooking chamber. A fan is used to ensure equal distribution of the signal throughout the oven; the microwave signal bounces off the rotating blades in all directions to equalize energy in all parts of the chamber. A mechanical or electronic (microprocessor-based) timer determines the presettable cooking interval.

In this experiment, you will open up a microwave oven, identify its major components, and verify its operation.

PROCEDURE

1. Open the housing of the microwave oven to gain access to the components and circuitry.
2. Locate the magnetron tube. Remove it from its socket and examine it. What are the main electrical connections to it?
3. Examine the waveguide assembly that transports the signal to the cooking chamber.
4. Locate the fan that distributes the signal.
5. Reinstall the magnetron and any related hardware.
6. Besides the magnetron, fan, and timer, what is the other major assembly in the oven?
7. With the microwave oven reassembled, test it by placing a neon bulb in the cooking chamber. Start the oven. The neon bulb should glow if microwave energy is present.

CHAPTER | 11

Introduction to Satellite Communications

ACTIVITY 11-1
TEST: INTRODUCTION TO
SATELLITE COMMUNICATIONS

Read the question and write the answer on a separate sheet of paper.

1. The period of a geosynchronous satellite is _____ h.
2. The distance of a geosynchronous satellite from earth is about _____ m.
3. The part of the earth that a geosynchronous satellite orbits around is the _____.
4. The locations of a satellite closest to and farthest from the earth while in an elliptical orbit are called the _____ and _____.
5. How are satellites moved to adjust their attitudes in orbit? _____
6. Satellites are positioned and held in orbit by either _____ or _____ stabilization methods.
7. Geostationary satellites are located by expressing their _____.
8. To aim a ground station antenna at a satellite, the _____ and _____ angles must be known.
9. The basic function of a communications satellite is to serve as a(n) _____.
10. The signal paths to and from a satellite are called the _____ and _____.
11. The transmitter-receiver unit in a communications satellite is known as a(n) _____.
12. Two popular satellite operating frequencies are the _____ and _____ bands.
13. The typical bandwidth of a satellite receiver is _____ MHz.
14. Name the main circuits of a transponder. _____
15. (True of False) Satellites can handle either digital or analog signals.

16. Each channel in a satellite requires a separate _____.
17. Explain how frequency reuse doubles channel capacity. _____

18. Explain how spatial isolation increases channel capacity. _____

19. Name the six main subsystems of a satellite. _____

20. The main battery power supply in a satellite is recharged by _____.

21. The main power amplifier in a transponder is usually a(n) _____.

22. In a regenerative repeater, the intelligence signal is _____ and used to _____ the carrier.

23. Bandpass filters in a transponder are usually of the _____ type.

24. The TTC subsystem allows a ground station to _____ and _____ the satellite.

25. (True or False) A satellite or ground station antenna cannot be used simultaneously for transmission and reception. _____

26. Typical IF frequencies in a dual conversion transponder are _____ and _____ MHz.

27. Name the six basic subsystems in a ground station. _____

28. The subsystems that communicate with earth-based telephones or radios are referred to as _____.

29. High-power earth stations use _____ or _____ amplifiers.

30. Frequency translations in an earth station are handled by _____ and _____.

31. A device that allows multiple transmitters or receivers to share an antenna is called a(n) _____.

32. Besides communications, satellites are widely used for _____.

ACTIVITY 11-2
PROJECT: SATELLITE GROUND
STATION ANTENNA

PURPOSE

To examine the parabolic dish antenna used in a satellite earth station to determine its characteristics and orientation.

MATERIALS

A nearby, accessible satellite dish antenna

Measuring tape (up to 15 or 20 ft)

Compass

INTRODUCTION

A ground station or earth station that accesses a communication satellite may be a receive-only (RO) station or one that transmits and receives (T/R). The RO stations are far more common and include business and industrial firms and educational institutions. Cable TV companies are receive-only stations. Consumer home satellite receivers are RO units. Earth stations that transmit are less common and include a variety of sources including publishers, TV stations, and telephone companies.

All those stations have one thing in common: They use a high-gain directional parabolic dish antenna pointed at a geostationary satellite.

The antenna, in fact, is the key link between the receiver and the satellite 22,300 mi away.

Because the antenna is so important, you should become more familiar with it, and this project gives you that opportunity. You will pay a prearranged visit to the site of a satellite antenna, inspect the antenna carefully, and determine some important information about it.

PROCEDURE

1. Go to the antenna site designated by your instructor.
2. Ask your instructor what band the ground station is using (C, Ku, etc.). Determine the frequency range. Record both the band and frequency range.
3. Measure the diameter of the dish with a tape measure. Be careful that you do not damage or disorient the antenna. An approximate measurement is good enough for this activity. Record your measurement.
4. Calculate and record the gain and beam width of the antenna at the lowest frequency in the designated band.
5. Note the orientation of the antenna. In what direction is it pointed? Explain.
6. Examine the feed system of the antenna. The antenna is usually a horn. Identify it and the transmission line.
   a. Which type of feed is used? Standard or Cassegrain?
   b. What type of transmission line is used? Waveguide or coax? Obtain measurements and types.
7. Note the mechanism for adjusting the orientation of the antenna; every antenna has a mechanical adjustment system. It may be manual and fixed, or it could be a motorized variable system. Inspect the antenna and record what you see.

CHAPTER | 12

Data Communications

ACTIVITY 12-1
TEST: DATA COMMUNICATIONS

Read the question and write the answer on a separate sheet of paper.

1. (True or False) In data communications, most binary transmissions are serial. _____

2. Name three well-known alphanumeric codes. _____

3. A digital circuit commonly used in serial-parallel/parallel-serial conversions is the _____.

4. The ASCII code for J is _____.

5. The ASCII code 1101101 means _____.

6. (True or False) "Bit rate" and "Baud rate" mean the same thing. _____

7. (True or False) The bit rate can be higher than the baud rate. _____

8. Start and stop bits are used in which kind of transmission? _____
   a. Synchronous
   b. Asynchronous

9. Which form of transmission is fastest and most efficient in communicating large volumes of data? _____
   a. Synchronous
   b. Asynchronous

10. State Hartley's law. _____

11. A twisted-pair cable has a bandwidth of 100 kHz. The maximum channel capacity is _____ bits/s.

12. A six-level code is used on a cable with a bandwidth of 15 kHz. The channel capacity is _____ bits/s.

13. The channel capacity on a 5-kHz bandwidth line with a 10-dB signal-to-noise ratio is _____ bits/s.

14. (True or False) With multilevel coding, the bit rate is higher than the baud rate. _____

15. A 9600 bit/s rate is desired. What minimum bandwidth is required if eight-level coding is used? _____

16. The device that makes binary signals compatible with the telephone lines is the _____.

17. The output of the modulator and the input to the demodulator in a modem is _____.
    a. Digital
    b. Analog

18. State the purpose of a UART IC. _____

19. The digital modulation scheme in which different frequency sine waves represent binary 0 and binary 1 is called _____.

20. What is a full-duplex modem? _____

21. A balanced modulator is used to generate _____.
    a. FSK
    b. PSK

22. A carrier recovery circuit is needed with? _____
    a. DPSK
    b. QPSK

23. In QPSK, _____ bits are coded per symbol.

24. A combination of QPSK and AM is known as _____.

25. A protocol defines _____ and _____ to be used in communications.

26. Special _____ are used at the beginning and end of a block of synchronous data for handshaking purposes.

27. Noise causes bit _____ in digital data transmission.

28. The logic circuit used in generating or checking parity is the _____.

29. Write the correct parity bit for the following words:
    Odd   1000101_____
    Even   0110100_____

30. Another name for vertical redundancy check is _____.

31. The _____ is generated by exclusive ORing all bits in all characters transmitted and appending the result to the end of the message.

32. Explain the process of generating the CRC character. _____

33. Each user station in a communications network is referred to as a(n) _____.

34. An interconnection of fewer than 100 users over a short distance for computer communications is called a(n) _____.

35. Define the terms "bridge" and "gateway." _____

36. Name the three basic LAN topologies. _____

37. An example of a star LAN is the widely used _____.

38. Name two popular types of cabling used in LANs. _____

39. LANs using modulation techniques are referred to as _____.
    a. Broadband
    b. Baseband

40. (True or False) Fiber-optic cable is used in LANs. _____

ACTIVITY 12-2
LAB EXPERIMENT:
SERIAL DIGITAL DATA

PURPOSE

To demonstrate one way to develop serial binary data and how to interpret an oscilloscope display of that data.

MATERIALS

Qty.

1	Dual-trace oscilloscope
2	4035B 4-bit CMOS shift register ICs
1	555 timer IC
1	10-bit DIP switch
11	4.7-kΩ resistors
1	22-kΩ resistor
1	0.01-μF capacitor
1	100-μF electrolytic capacitor

INTRODUCTION

In virtually all data communications applications, binary data is transmitted serially. That is, rather than transmit all bits of a binary word simultaneously or in parallel, the data is transmitted serially one bit at a time. The reason for this in communications applications is obvious. To transmit parallel data, multiple paths or channels, one per bit, are required. In most communications systems, only a single channel is available. For that reason, multiple binary words are transmitted one after another one bit at a time.

On the other hand, binary data is usually generated and manipulated by parallel methods in the electronic equipment not used for communications. That is particularly true in computers. For that reason, the processes of parallel-to-serial and serial-to-parallel data conversions are important.

The circuit most commonly used to perform parallel-to-serial and serial-to-parallel data conversions is the shift register. A shift register is a digital circuit made up by cascading D or JK flip-flops. To translate a parallel binary number into a serial number, the parallel number is loaded into the shift register flip-flops, and then a clock signal is applied simultaneously to all of the flip-flops. The resulting data is shifted one bit to the right (or to the left in some cases), thereby generating a serial train of data at the output flip-flop.

The process of parallel-to-serial data conversion is illustrated in Fig. 12-1. An 8-bit parallel binary number is loaded into the flip-flops of an 8-bit shift register, and then clock pulses are applied to all flip-flops simultaneously. When the first clock pulse occurs, all of the bits in the binary number are shifted one position to the right. The bit in the right-hand flip-flop is shifted out. That bit will then be applied to some other circuit or transmitted over the communications channel. An example is the modulator in a modem for transmission over the telephone lines.

When the second clock pulse occurs, all of the bits are again shifted one position to the right. The bit in the rightmost position is shifted out. With each succeeding clock pulse, the bits in the shift register are moved to the right as the number is completely shifted out. By the time the eighth clock pulse occurs, all bits will have been shifted. Note that the

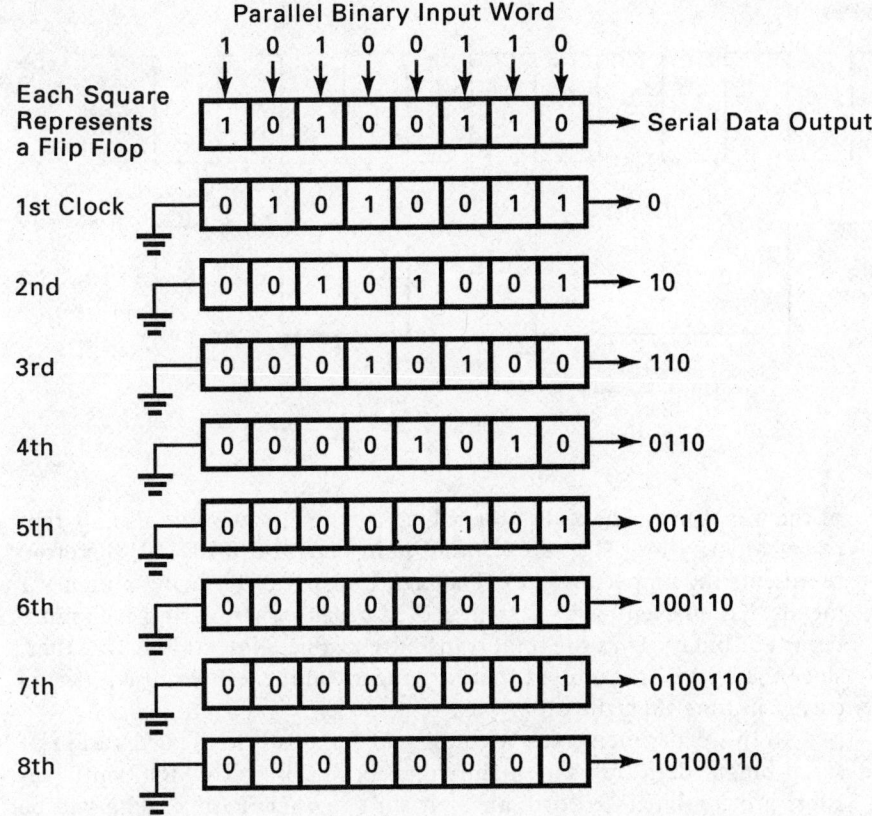

Parallel Binary Input Word

1 0 1 0 0 1 1 0

Each Square Represents a Flip Flop

Clock		Flip Flop Contents	Serial Data Output
	1 0 1 0 0 1 1 0	→ Serial Data Output
1st Clock	0 1 0 1 0 0 1 1	→ 0
2nd	0 0 1 0 1 0 0 1	→ 10
3rd	0 0 0 1 0 1 0 0	→ 110
4th	0 0 0 0 1 0 1 0	→ 0110
5th	0 0 0 0 0 1 0 1	→ 00110
6th	0 0 0 0 0 0 1 0	→ 100110
7th	0 0 0 0 0 0 0 1	→ 0100110
8th	0 0 0 0 0 0 0 0	→ 10100110

Fig. 12-1 Shift register used for parallel-to-serial data conversion.

input to the left-hand flip-flop of the shift register is connected to ground. That represents a binary 0. As the data is shifted out to the right, binary 0s are shifted into the flip-flops.

If you were to observe the serial data output of the shift register of Fig. 12-1, you would see a signal like that shown in Fig. 12-2 on the next page. The upper waveform is the clock, a series of periodic pulses. The clock pulses are numbered to correspond to the clock signals designated in Fig. 12-1.

The serial data shows the voltage waveform at the output of the right-hand flip-flop in the shift register. After the first positive-going transition of clock pulse 1, a binary 0 output appears. When the second positive-going transition of clock pulse 2 occurs, a binary 1 appears at the output. Note the sequence of the bits in Fig. 12-2 and compare it to the sequence of the bits in Fig. 12-1. They appear to be reversed. The reason is that an oscilloscope displays voltage levels with respect to time, with time proceeding from left to right. You will see the binary data displayed from left to right on the screen as it is shifted out.

An important consideration in transmitting serial data is which bit of a binary word is transmitted first, the least significant bit (LSB) or the most significant bit (MSB). It is important to know that so you can interpret the data after it has been transmitted. In data communications systems, various schemes are used when either the MSB or LSB is transmitted first. You simply must become familiar with the protocols and the system specifications to determine this information.

For example, when ASCII data is transmitted, the LSB is usually transmitted first. Recall that the ASCII system represents letters of the alphabet, numbers, punctuation symbols, and other characters as 7-bit binary numbers. The serial word in Fig. 12-2 might represent the ASCII 0100110. The LSB in that case would be the zero at the left-hand side

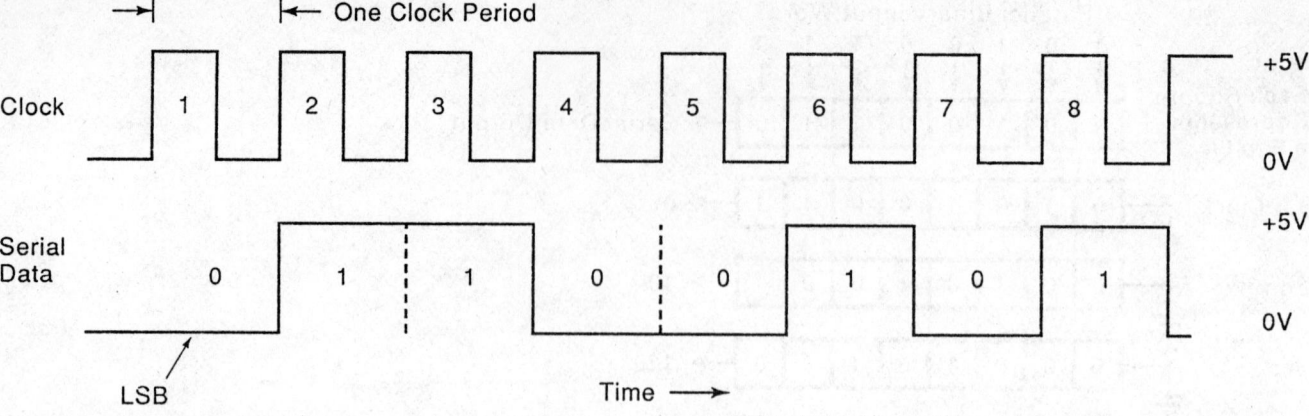

Clock

Serial
Data

LSB

Time →

Fig. 12-2 The serial binary waveform.

of the waveform. The eighth bit, which is a binary 1, would usually represent a parity bit. The ASCII number in Fig. 12-2, 0100110, therefore represents the ampersand (&). The eighth or parity bit being a binary 1 means that an even-parity system is used because there are an even number (4) of binary 1s in the total transmitted value. Note in Fig. 12-2 that, since the LSB is transmitted first, it appears on the left because it occurs earlier in time than the other bits.

In this experiment, you are going to demonstrate the generation of serial binary data by using a shift register. Two 4-bit CMOS shift registers are combined to form an 8-bit shift register. Parallel data will be loaded into this register by way of a set of DIP switches. An oscillator made up of a 555 timer IC will be used to produce the clock pulses to cause the data to be shifted out. In order that you can view the data on an oscilloscope, the serial data output from the shift register is connected back to the serial input of the shift register. In other words, as the data is shifted out, it is also shifted back into the same register. That allows the word to be shifted out and then reloaded again and again for repetitive viewing on an oscilloscope. By loading in different binary values and displaying the clock pulses and serial data output simultaneously, you will be able to determine the relationship between the parallel input bits and the serial output bits.

PROCEDURE

1. Construct the circuit shown in Fig. 12-3. Because of the complexity of the circuit, be extremely careful in making the interconnections to avoid wiring errors.

   The pin-out diagram for the 4035B shift register is shown in Fig. 12-4 on page 92; that is how the pins on the 16-pin DIP are numbered when you view the IC from the top. The P1 through P4 pins accept a 4-bit parallel binary input word. The individual flip-flop outputs are designated Q_1 through Q_4; Q_1 is the flip-flop nearest the serial input. The serial input is applied to the J and K NOT inputs on pins 3 and 4. Usually these two pins are tied together. Pin 6 accepts the CLOCK input. Pin 5 is the reset (RST) line. Bringing this line high or to binary 1 causes all flip-flops to be simultaneously reset. The LOAD input at pin 7 is used to cause the parallel binary inputs to be loaded ino the shift register flip-flops. Normally this pin is held low, or binary 0, for normal shifting operations. To load the parallel word, pin 7 is brought high, or binary 1. The occurrence of the first clock pulse

90

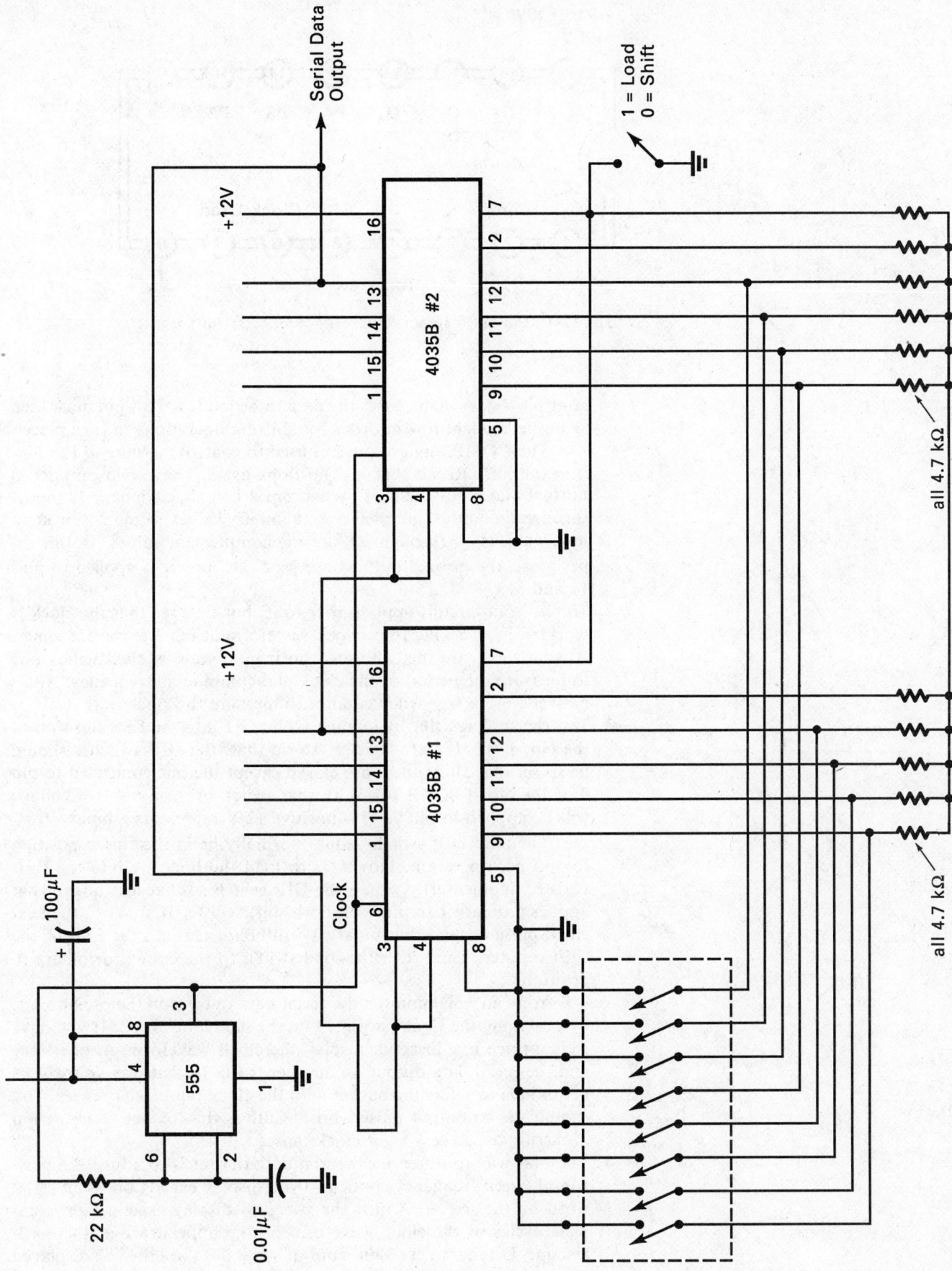

Fig. 12-3 An 8-bit parallel to serial data converter.

91

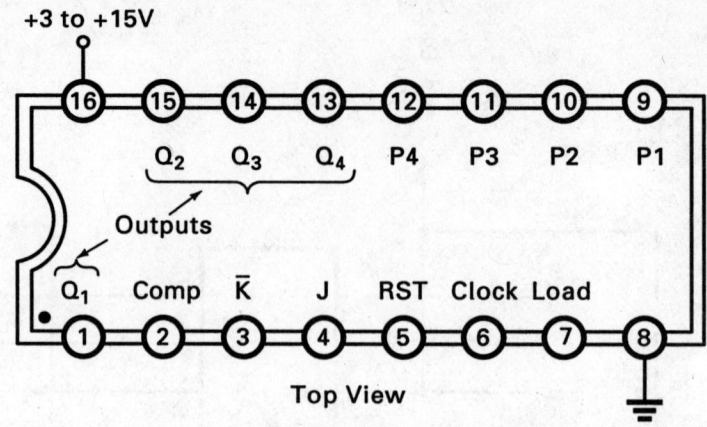

+3 to +15V

Top View

Fig. 12-4 Pin-out diagram of 4035B 4-bit CMOS shift register.

after pin 7 goes high causes the data to be loaded. This pin must then again be brought low in order for shifting operations to take place.

The COMP input at pin 2 is used to control the state of the flip-flop outputs. Recall that all flip-flops usually have two outputs: a normal and a complement. When pin 2 is high, or binary 1, the Q_1 through Q_4 outputs are their normal binary values. If pin 2 is brought low, outputs Q_1 through Q_4 are the complement values. In this experiment, the normal outputs are used. DC power is applied to pins 16 and 8.

2. To test your circuit, apply power to it, but first see that the clock is working by checking for a clock waveform at pin 3 at the 555 timer. At that time, use the calibrated horizontal scale of the oscilloscope to measure the period of the clock and compute its frequency. Alternatively, use a frequency counter to measure the clock output.

3. Test the shift register by loading a binary 1 into the LSB flip-flop of the No. 1 4035B shift register. To do that, the DIP switches should be so set that all of them are closed except the one connected to pin 9 of the No. 1 4035B IC. With that switch open, a positive voltage will be applied to pin 9. This positive 12 V represents a binary 1.

The load DIP switch should normally be in the closed position for proper operation. However, to load the binary 1 into the shift register, momentarily switch the DIP switch so that it is open. That applies a binary 1 to pin 7 on both shift register ICs. When the next clock pulse occurs, that binary 1 will be loaded into the LSB of the shift register. Then reset the load switch to the closed, or binary 0, position.

You can now monitor the serial data output on the oscilloscope by observing the signal at pin 13 on the No. 2 4035B IC. If you have a dual-trace oscilloscope, display the clock waveform on one trace and the serial data output on another trace. In that way you will be better able to relate the output with the clock pulse occurrences. You should see an output pulse whose width is that of one clock period occurring once every eight clock pulses.

4. The best way to observe the serial data output is to adjust the horizontal sweep frequency control so you display exactly one 8-bit serial word on the screen. Adjust the sweep to display exactly eight complete cycles of the clock waveform on the upper trace of the oscilloscope. One complete 8-bit word of serial data will then be displayed on the lower trace.

5. To see the variation in the output waveform based upon the actual

value loaded into the shift register, try loading in different binary numbers and observing the binary output. The following are some examples:

10101010

00110011

11101110

00001111

In each case, set the DIP switches to apply the number to the shift register. The LSB shown above will be the value loaded into what we are calling the LSB of the shift register, which in this case is input pin 9 on the No. 1 4035B IC. The MSB then is pin 12 on the No. 2 4035B IC. Once the DIP switches have been set to the desired number, momentarily apply a binary 1 to pin 7 of the ICs with the load DIP switch. You should immediately note a change in the output pattern on the oscilloscope screen. Be sure that you can relate the binary numbers above to the corresponding oscilloscope patterns that they create.

6. Based upon the clock frequency measured in the first step, how long does it take to transmit a single 8-bit binary word?

7. Draw a diagram similar to the one in Fig. 12-1 that illustrates how a serial binary data word is captured by shifting it into a shift register. Illustrate the process in the same number of steps to show how serial-to-parallel data transfer occurs. Use the ASCII code for the letter M with odd parity.

ACTIVITY 12-3
LAB EXPERIMENT:
FREQUENCY SHIFT KEYING

PURPOSE

To demonstrate the generation of a frequency shift keying signal.

MATERIALS

Qty.
1   Dual-trace oscilloscope
1   Square wave signal source
1   2206 function generator IC
1   150-Ω resistor
2   4.7-Ω resistors
1   22-kΩ resistor
1   47-kΩ resistor
1   56-kΩ resistor
1   0.001-μF capacitor
1   1-μF capacitor
1   10-μF capacitor

INTRODUCTION

Frequency shift keying is a modulation technique used to transmit binary data over analog communications channels such as the telephone lines. The binary data operates an FSK modulator that translates the binary signals into two different tones or sine waves. The frequencies chosen depend upon the bandwidth of the analog channel as well as the speed of the binary data. Because the telephone lines have a very narrow bandwidth (approximately 3 kHz), low frequencies are used. For example, in low-frequency modems, the frequencies of 1070 and 1270 Hz are used to represent the binary values 0 and 1, respectively. In full-duplex systems, the tones 2025 and 2225 Hz are also used to represent binary 0 and 1, respectively. When a binary 0 occurs, the low-frequency tone is generated; when a binary 1 occurs, the high-frequency tone is generated. Other schemes have also been used. Because such low frequencies are used, the data rate is restricted; the maximum with such a system is typically in the 300 bits/s region. For higher data rates, other types of modulation such as PSK and QAM are used.

In this experiment you will demonstrate the generation of FSK and observe the output signal. You will use the 2206 function generator IC which you used in preceding experiments. This IC was specifically designed for FSK applications. The serial binary input will be simulated by a square wave.

PROCEDURE

1. Construct the circuit shown in Fig. 12-5. You may have retained the 2206 function generator circuit used in many preceding experiments. If so, modify it so that it appears as shown in Fig. 12-5.
2. Apply a square wave to pin 9 of the 2206 IC; it will simulate a serial binary input. The square wave should have an amplitude that switches between 0 and approximately +5 V. Any TTL data source will be suitable. Set the frequency to approximately 1.5 kHz.
3. Apply power to the 2206. Monitor the FSK output at pin 2. Because the input signal repeatedly switches the output between two different sine wave frequencies, scope triggering will be somewhat erratic. However, through a combination of scope triggering and horizontal sweep frequency control adjustments, you should be able to stabilize the waveform and see the two different frequencies as the input signal switches between binary 0 and binary 1. Use the variable sweep frequency and trigger level controls to stabilize the waveform. If you

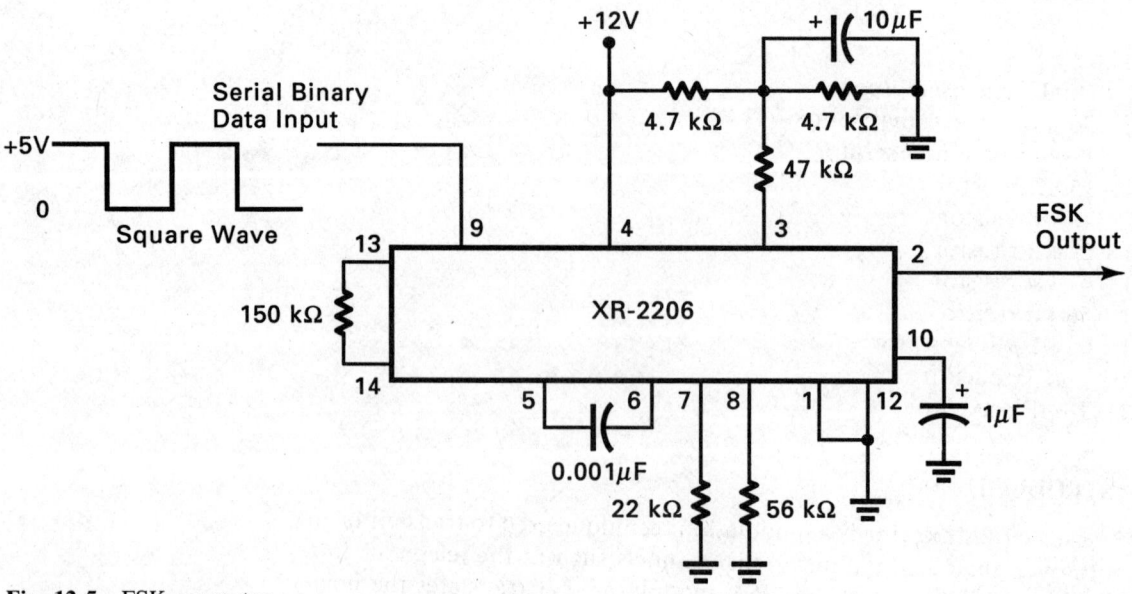

Fig. 12-5 FSK generator.

have a dual-trace oscilloscope, you can display the binary input signal on the upper trace and the FSK output signal on the lower trace.

4. Determine which binary input state produces which output frequency. One way to do that is to remove the square wave from pin 9 and then ground pin 9 to provide a fixed binary 0 input. Measure the 2206 output frequency at pin 2 with a frequency counter or on the oscilloscope. Apply +5 V dc to pin 9 and again measure the 2206 output at pin 2. That will simulate a binary 1 input. Record the information you have collected.

5. The frequencies of the output tones produced by the 2206 arc determined by resistor and capacitor values; specifically, the 0.001-μF capacitor connected between pins 5 and 6 and the 22- and 56-kΩ resistors connected to pins 7 and 8 in Fig. 12-5. Knowing that, what steps would you take to reverse the frequency tones you determined in step 4? That is, how would you change the tone obtained with binary 0 so that it is produced when a binary 1 occurs, and vice versa?

6. Suggest a type of circuit that might be used to demodulate an FSK signal.

CHAPTER | 13

Fiber-Optic Communications

ACTIVITY 13-1
TEST: FIBER-OPTIC COMMUNICATIONS

Read the question and write the answer on a separate sheet of paper.

1. In fiber-optic communications, _____ is used as the carrier for intelligence signals.

2. Fiber-optic cable is made of _____ or _____.

3. The light source in a fiber-optic system is either a(n) _____ or a(n) _____.

4. The signals transmitted over a fiber-optic cable are usually _____.
   a. Analog
   b. Digital

5. The main advantage of fiber-optic cable over conventional wire cables is _____.

6. List seven other benefits of fiber-optic cable over electrical cables.

7. (True or False) Fiber-optic cables are nonconducting. _____

8. (True or False) Unlike a radio signal, light is not an electromagnetic signal. _____

9. The wavelength of visible light is _____ to _____.

10. The speed of light is _____.

11. The speed of light in a fiber-optic cable compared to that in free space is _____.
    a. More
    b. Less
    c. The same

12. Nonvisible light known as _____ is often used in fiber-optic cables.

13. The frequency of infrared light compared to that of visible light is _____.
    a. Less
    b. More
    c. The same

14. The path of a light beam is a(n) _____.

15. The phenomenon of refraction causes light waves to _____ when passing from one medium to another.

16. If a light beam strikes a glass surface at an angle greater than the critical angle, it is _____.
    a. Absorbed   c. Reflected
    b. Refracted   d. Attenuated

17. Light passes through a fiber-optic cable by being repeatedly _____ from the sides.

18. Name the three basic types of fiber-optic cable. _____

19. Which type of cable has the least attenuation? _____
    a. Plastic
    b. Glass

20. Modal dispersion causes pulse _____, which reduces the _____ of data transmission.

21. A fiber-optic cable ½ mi long with an attenuation of 8 dB/km is spliced to a 1¼-mi-long cable with an attenuation of 11 dB/km. The total attenuation is _____ dB.

22. To avoid excessive light loss when splicing fiber-optic cables, the cable ends must be carefully _____.

23. (True or False) Like electrical cables, fiber-optic cables are linked by connectors. _____

24. A commonly used LED light source frequency is _____ because fiber-optic cable offers minimum attenuation at that frequency.

25. High-speed data is best transmitted with a(n) _____.
    a. LED
    b. Laser diode

26. The light source in a low-speed short-distance fiber-optic system is a(n) _____.

27. For proper operation, a photo diode must be _____.
    a. Reverse-biased
    b. Forward-biased

28. Name two fast photodiodes that are used in fiber-optic systems.

29. A fiber-optic system has an 85-Mbits · km/s rating. The rating of a 15-km cable is _____.

30. The maximum distance possible between repeaters in today's fiber-optic systems is _____ km.

ACTIVITY 13-2
LAB EXPERIMENT: OPTICAL
FIBER DATA TRANSMISSION

PURPOSE

To demonstrate the transmission of binary data over a simple fiber-optic link.

MATERIALS

Qty.
1 Oscilloscope
1 Function generator (TTL square wave)
(*continued*)

1 Fiber-optic cable (3 ft)
1 Motorola MFOE71 LED with housing and connector
1 Motorola MFOD72 photo transistor with housing and connector
2 NPN transistors (2N3904, 2N4401, or similar)
1 100-Ω resistor
1 1-kΩ resistor
1 3.9-kΩ resistor
1 33-kΩ resistor
1 220-kΩ resistor

INTRODUCTION

More and more data communication is being carried out over optical fiber cables rather than via twisted-pair, coax, or other wire conductor cables. That includes large communications systems like the telephone network as well as smaller systems such as local area networks. The high speed, high capacity, wide bandwidth, low noise, high security, and other characteristics give fiber-optic equipment major advantages over older and more conventional data communications techniques. If you plan to work in the electronics communications field, you will most surely encounter fiber-optic equipment. The purpose of this experiment is to familiarize you with the basic elements of all fiber-optic systems.

First you will examine the fiber-optic cable itself and note its construction and characteristics. You will learn how to cut the end of a cable and mount it in connectors. Next you will construct an LED transmitter and send light pulses down the cable. Finally, you will build a receiver with a phototransistor and demonstrate the reception of the light pulses. This experiment will give you a good feel for the hardware and operation of typical fiber-optic equipment.

PROCEDURE

1. Locate the fiber-optic cable. Use a wire stripper to remove about $\frac{1}{8}$ in. of outer insulation to expose the fiber element itself. Note the clear plastic fiber. Look at it under a magnifying glass if one is available.
2. Point one end of the cable at a strong light source; it can be a window (daylight, of course), a lamp, an overhead light fixture, a flashlight, or whatever is convenient. Observe the other end of the cable, and you will see the light being transmitted. Put your finger over the end of the cable pointed at the light and note how the light at the other end disappears. Point the end of the cable at several light sources with different intensities and see how the light at the other end varies. Experiment further as desired.
3. Construct the transmitter circuit shown in Fig. 13-1. The physical connections to the transistor and LED also are given. Do *not* connect the cable to the LED yet.
4. Connect a TTL square wave from a function generator to the input of the transmitter and set the transmitter frequency to about 1 Hz. Apply power to the transmitter circuit and look inside the hole where the fiber cable goes. You should see the faint red glow of the LED switching off and on at a 1-Hz rate.
5. Turn off the power and wire the receiver circuit as in Fig. 13-2 on page 100.
6. Prepare the fiber-optic cable by stripping about $\frac{1}{8}$ in. of outer sheath from each end. Use a sharp razor blade or X-Acto knife to cut the

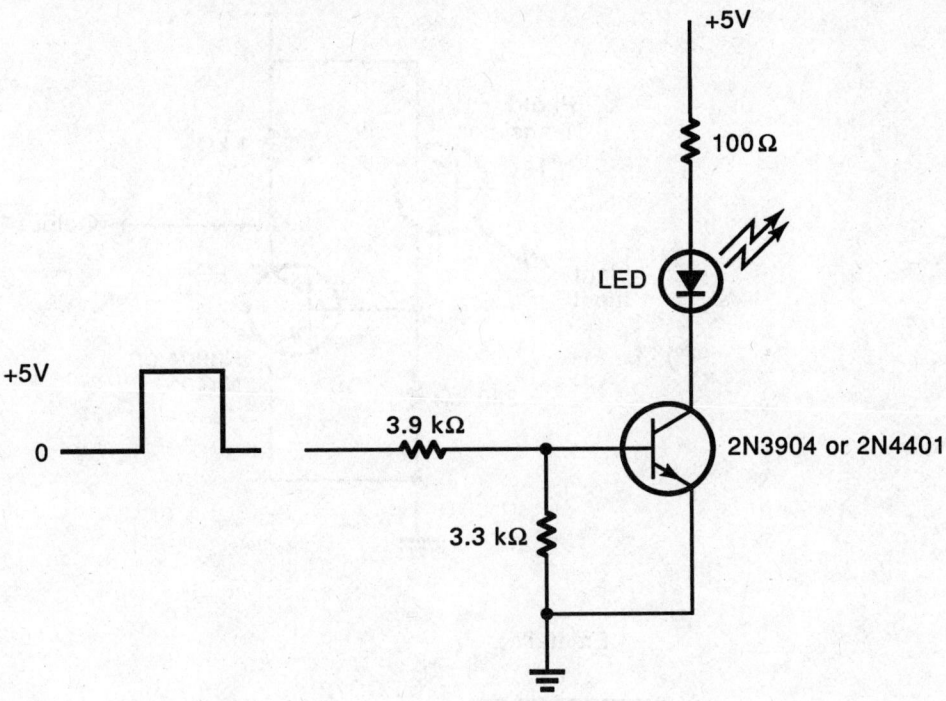

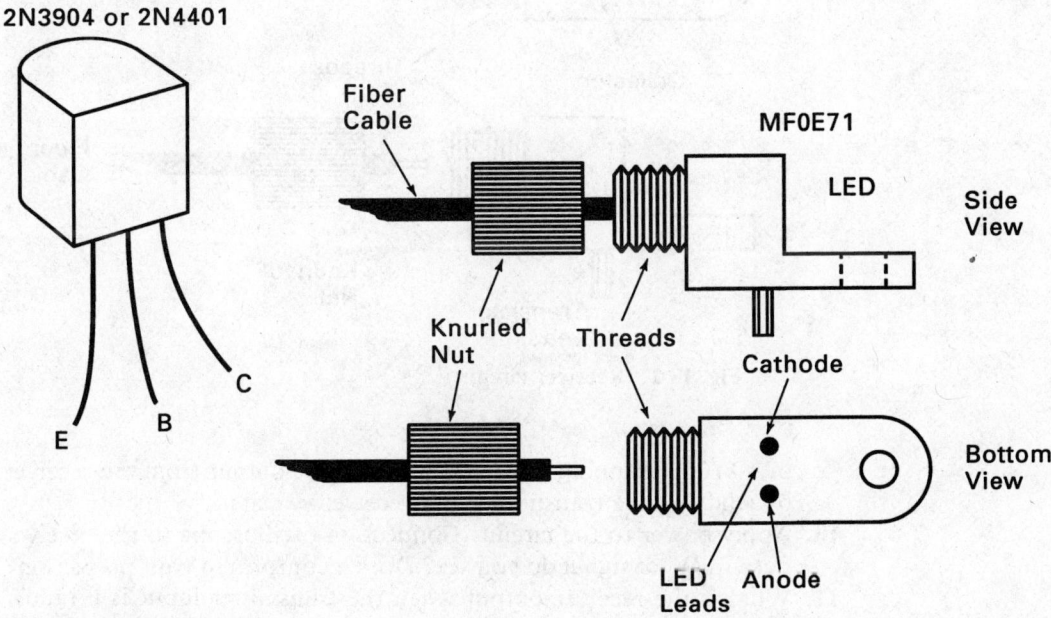

Fig. 13-1 LED transmitter.

end of the fiber clean and square. Rotate the cable as you cut so that you get a flat, even end surface.

7. Remove the knurled nut from the LED housing and slip it over one end of the fiber-optic cable. Then insert the fiber into the LED housing as Fig. 13-1 shows. Push the fiber cable all the way in and then tighten the nut.

8. Repeat the above procedure for the other end of the cable and install the cable in the photo transistor assembly.

9. Set the function generator to supply a square wave with a frequency

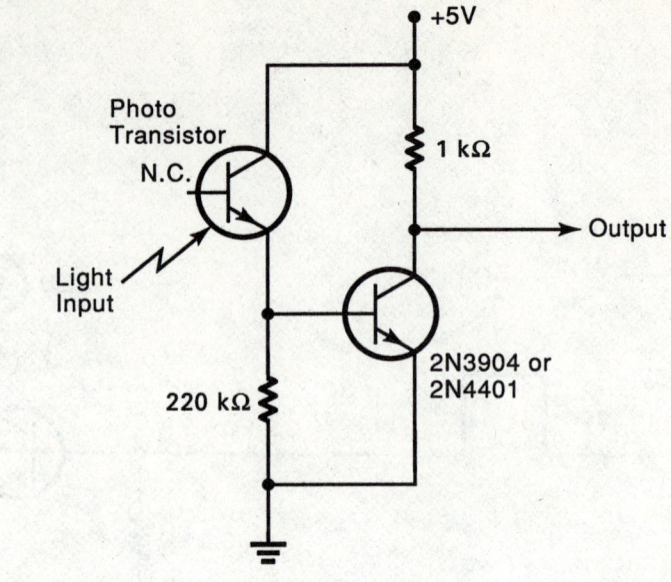

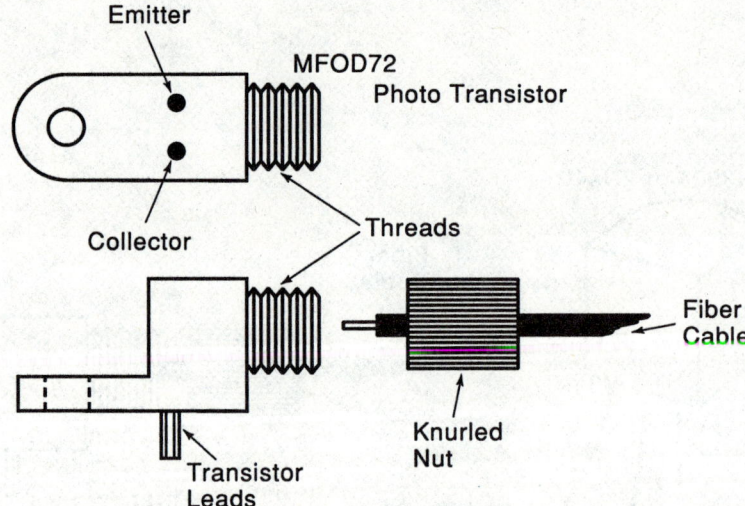

Fig. 13-2 Receiver circuit.

of 1 kHz. Assuming that input, predict the output from the receiver by studying the transmitter and receiver circuits.

10. Apply power to the circuit. Connect an oscilloscope to the receiver output. What signal do you see? Does it conform to your prediction?

11. What is the receiver output when the transmitter input is 0 (gnd)? What is it when the receiver input is 1 (+5 V)? Record both results.

12. While observing the receiver output, begin increasing the function generator frequency slowly beyond 1 kHz. Describe what happens. Explain why. What is the maximum frequency before the output begins to decrease in amplitude?

13. Verify that it is the light over the fiber-optic cable that is operating the receiver. Disconnect the cable from the transmitter end while observing the pulse output at the receiver. Describe what you see. How close does the cable have to be to the LED to see an output?

14. Can analog signals be transmitted over a fiber-optic link?

CHAPTER | 14

Modern Communications Applications

ACTIVITY 14-1
TEST: MODERN COMMUNICATIONS
APPLICATIONS

Read the question and write the answer on a separate sheet of paper.

1. (True or False) Fax machines transmit sound as well as pictures.
   _____

2. Most fax transmissions take place via _____.
   a. Radio
   b. Satellite
   c. Telephone lines

3. Pictures are converted to electrical signals in a fax machine by _____ the picture with a light-sensitive device.

4. The light-sensitive device used in most fax machines is the _____.

5. List the four types of fax transmission systems and the types of modulation and resolution. _____

6. The most widely used fax format is _____.

7. The technique used to speed up transmission in most modern fax machines is _____.

8. The type of printer used in most fax machines is _____.
   a. Impact
   b. Thermal
   c. Laser

9. Each cell in a cellular telephone system consists of a(n) _____ and _____ for communicating with mobile units.

10. Each cell communicates with the _____, which is the link to the main telephone system.

11. What technique permits many more cells and frequency channels to be available in a given spectrum space? _____

12. The frequency range of the cellular telephone system is _____ to _____ MHz.

13. Cellular telephones operate _____.
    a. Half duplex
    b. Full duplex

14. Cellular telephones use _____ modulation.

15. Cellular channels are spaced _____.

16. The spacing between cellular transmit and receive channels is _____.

17. The time for transmission and reception of a radar pulse is 56 μs. The distance to the target is _____ nautical miles.

18. A highly directional antenna permits a radar to determine which characteristic of a target? _____
    a. Range
    b. Speed
    c. Bearing
    d. Altitude

19. Name the two types of radars. _____

20. Name the effect that permits measuring the speed of a moving object by radar. _____

21. The output display of a radar set is usually a(n) _____.

22. List four radar applications. _____

ACTIVITY 14-2
PROJECT: FACSIMILE
FAMILIARIZATION

PURPOSE

To become familiar with the operation and construction of a modern fax machine.

MATERIALS

Operating fax machine (suitable for disassembly)
Manufacturer's service and operating manuals

INTRODUCTION

The fax machine, along with the telephone, the copier and the PC, has become one of the most widely used business tools. It is one of those high-tech machines that people wonder how they ever got along without. Once its speed and convenience are experienced, individuals find it hard to go back to the slower, less-convenient alternatives. In this activity you will gain some basic familiarity with this popular machine.

PROCEDURE

1. Disassemble the fax machine supplied by your instructor so that the interior circuits and mechanisms are accessible. Examine the unit carefully and identify the following major elements:
    a. Power supply
    b. Telephone
    c. Scanning optics and paper-handling mechanism
    d. CCD sensor

e. Modem circuits

f. Printer

If available to you, use the manufacturer's manuals to help you.

2. Use this fax machine or another one to send a document to someone else. That will help you become familiar with the operating procedures.

3. Use this fax machine or another one to receive a fax document. Observe the entire procedure from answering the call to receiving the final printed document.

4. Examine the received document and study the clarity of reproduction. Note the limited resolution on small print and other fine details and the ''jaggies'' on long diagonal straight edges. Examine the quality of any photographs sent.

5. Make a phone call to a telephone number with a known fax machine. Listen to the tones and describe what you hear.

ACTIVITY 14-3
FIELD TRIP:
CELLULAR TELEPHONE

PURPOSE

To visit the mobile telephone switching office (MTSO) and/or a cell site to become familiar with the equipment.

INTRODUCTION

Most large and medium-size cities now have cellular telephone systems. The larger metropolitan areas authorize two companies to provide cellular service for the area. Each company will have a switching office (the mobile telephone switching office) with its computer, transmitting, and receiving equipment and the links to the telephone system. Each will also have multiple cell sites to serve the designated geographic area. The purpose of your field trip is to visit the MTSO and/or a cell site so you can see the equipment.

PROCEDURE

1. Go with your instructor and class to the MTSO and cell site at the time and place designated.

2. Ask to see a map of the system showing the MTSO and cell site locations. How many cells are used? What is the shape of the cell coverage?

3. Ask specifically to see the following equipment:
   a. Receivers
   b. Transmitters
   c. Power supplies including backup
   d. Control and switching
   e. Wire line or microwave links to cell sites
   f. Antennas and related couplers (diplexers) and the transmission line

4. Ask who the manufacturers of the equipment are.